After a successful career in journalism, Mike Tomkies returned to his childhood love of nature and spent over thirty years living in remote places in the Scottish Highlands, Canada and Spain. During this time, he wrote a number of best-selling books, including *Alone in the Wilderness, A Last Wild Place, Out of the Wild, On Wing and Wild Water, Wildcat Haven* and *Moobli*. He died in 2016.

Between Earth and Paradise

An Island Life

MIKE TOMKIES

Foreword by Polly Pullar

This edition published in Great Britain in 2021
under licence from Whittle Publishing by
Birlinn Ltd
West Newington House
10 Newington Road
Edinburgh
EH9 1QS

www.birlinn.co.uk

ISBN: 978 1 78027 706 6

Copyright © The estate of Mike Tomkies
Foreword copyright © Polly Pullar, 2021
First published in 1981 by William Heinemann

Subsequently published by Whittles Publishing, Dunbeath in 2006

All rights reserved. No part of this publication may be
reproduced, stored, or transmitted in any form, or by any means,
electronic, mechanical or photocopying, recording or otherwise,
without the express written permission of the publisher.

British Library Cataloguing-in-Publication Data

A catalogue record for this book is available on
request from the British Library
Papers used by Birlinn are from well-managed
forests and other responsible sources

Typeset by Hewer Text UK, Edinburgh
Printed and bound by MBM Print SCS Ltd, Glasgow

MIX
Paper from
responsible sources
FSC® C117931

Contents

'In the time of middle earth, timeless aeons before the fall from grace, there was a place where men could pass quite freely between earth and paradise. So that men's minds should not be dazed by the shocking beauty of paradise, the place was made in the celestial image, a high place of ethereal beauty, masked in soft mist, earthly yet shot through with a spirit beyond the range of men's minds. So perfect it could not be other than the roots of heaven.'

– A legend told in the Allinson family of Westmorland

Foreword

Reading *Between Earth and Paradise* again 40 years after it was first published brings a lump to my throat. It had the same effect first time when I was in my youth and it was hot off the press. I wonder if I realised then that its depth and poignancy would become even more pertinent all these years later. Did I realise that in less than half a century I would witness the fruition of Mike Tomkies' oft-voiced fears for the fate of the natural world through the decline of myriad common species?

Tomkies' life as an accomplished naturalist and ambassador for nature began in early childhood. However, it was his sojourn on the tiny, beautiful island of Eilean Shona in Loch Moidart where the Scottish era of his isolated wilderness years took flight. Here he found his vocation. Before that, he had worked as a journalist amongst Hollywood's rich and famous, embroiled in a wildly flamboyant world dominated by drink and drugs, colourful love affairs and short-termism. Then, following a spell in the Canadian wilderness, he exchanged the glamour of Hollywood for a very different type of 'wild' and began his solitary existence as an accomplished backwoodsman, naturalist, writer and photographer. His prolific outpourings quickly gathered a large following and symbolised all that was wild and free.

Between Earth and Paradise transports the reader to Eilean Shona and to the trials he faced setting himself up in a long-abandoned croft. It is moving and inspirational. As the wind rampages around his draughty walls and rain batters down relentlessly he battles with himself, but finds deep resilience and draws on this regularly, for there is no time for self-pity. His days are filled with intrigue and discovery as he studies the minutiae of the rich wildlife that surrounds him, and he works ceaselessly to keep body and soul together.

There can be no doubt that his troubled childhood – he lost his mother when he was very young – and lack of human companionship led to occasional bouts of melancholy that challenged him as much as the savage storms that battered Eilean Shona. There are failing finances, self-doubts and lack of confidence too, as well as nostalgic thoughts of the woman he lost. As he watches animals pairing up, he cannot help wondering if he will find someone else with whom to share his life. But always he returns to nature, where he finds solace and meaning.

Though Tomkies' work is beautifully descriptive, he is not given to flowery verbosity – his style is captivating and extraordinarily tender. Information is seamlessly absorbed as if by osmosis through his accurate observations and his extraordinary attention to detail. There is humour too, particularly when revealing the characters of the animals and birds that share his life, from an injured orphan heron and a damaged herring gull to a hand-reared sparrowhawk that once set free returns for food, as well as spiders – one called Sarah – and the mice and moths that put in brief appearances. This is a man at one with all his subjects whether they be a dominating little cock chaffinch called Captain or a noble red deer stag, Sebastian, that filches his frugal hard-won vegetable harvest.

Heavily influenced by the eminent marine biologist and conservationist Rachel Carson, who prophesied our biodiversity crisis in her seminal work, *Silent Spring*, in 1962, Tomkies is painfully aware of the adverse effects humans have on the natural world. He writes frequently of his concerns regarding the damage done to the Highlands by the increase of visitors and the resulting shattering of peace. While clearing an area of rough hill where he is building himself a rustic cabin, he realises his presence may be disruptive to the wildlife and abandons the project entirely. Such actions reveal his extraordinary compassion. How many of us consider the needs of wildlife before our own, or take active steps to give something back to nature? Tomkies did. As he writes (pp. 84–85):

> The pattern had become crystal clear. Nature had been the one constant love throughout life. But it was no longer enough merely to enjoy the wilderness, use it as a retreat or for inspiration. I had to try and pay it back. It was then I made my one New Year resolution. From henceforth I would write only about nature and the last wild places and man's place in and influence upon them . . . Perhaps after 22 years of journalism, meeting man at his best and worst in half the western world, I could avoid the narrow, specialized naturalist's view. I knew now that only by giving myself totally to this new life, by trying to understand the magnificent Highland wilderness deeply, factually, and writing about it with reverence and love, had I the faintest chance of succeeding.

Mike Tomkies certainly succeeded. And, importantly, he made a difference, for during his 88 years he changed people's perception, raising an awareness of the continuing need to protect the natural world. This republication of *Between Earth and Paradise* is timely – and his message has become more urgent than ever.

Polly Pullar
Aberfeldy
March 2021

Introduction

It was a blustery, cool October day with heavy banks of black cloud scudding across a bright sky, delivering intermittent squalls of rain. I had always imagined that the last crossing of my beloved but treacherous Highland loch would be a painful and tear-filled occasion. Yet, now that I was actually embarked on that final short voyage, alone in my boat watching the mountains, the wooded glens and the steely grey choppy water slipping away behind me for the last time, I felt no melancholic emotion at all, only a dull anger. That, and resignation to the fact that my time was over here in this beautiful and terrifying remote spot. I was leaving Wildernesse, leaving Scotland, for good. There was nothing more I could do here. Nothing seemed to lie ahead but repetition, endless physical toil and mental atrophy. It was time for new horizons.

Perhaps it was my perception that was at fault. After so many years of battling home in winter storms, across frightening seas or up a pitch-dark lonely loch, enduring the countless testing chores necessary for survival in the wildest and harshest environment in Britain, without gas, electricity, television, phone, postal service, or even a road, perhaps I had grown apart from human society. Maybe I had isolated myself too much from the main ambitions of my fellow men in order to get close to the secret lives of the golden eagles, the rare divers, the pine-martens, the deer and the many other creatures with which I had shared my desolate 300-square-mile patch. But not only I had changed; people's attitudes had changed over the past decade. A growing awareness of the last wild places had brought scores of well-meaning seekers after solitude to trample down and disturb the delicate balance of nature in the Highlands, and greed had taken on the guise of progress. Much of the rare wildlife I had come to love was endangered now.

Fish farmers didn't seem to care that overproduction was threatening the nest sites of rare birds. Sheep farmers pocketed double subsidies and stocked the hillsides with their animals, threatening eagles with poison or harassing them at the nest after assuming, wrongly, that they took many live lambs.

Deer hunters were taking more and more of the finest stags, diluting the genetic pool, while the proliferating fenced conifer forests denied winter shelter and food to a burgeoning but weakened deer population. The fox-hunters were worse. This barbaric pursuit does not (contrary to the hunters' claims) 'control a pest', and when the masters of a nearby government-funded pack blew up with explosives the crag below an eagle's nest 'to free hounds trapped in a fissure', I felt I had encountered the ultimate in lunacy. Above all, noise and commotion had come to the once peaceful and silent glens.

I landed the boat on the far shore and began heaving my belongings up the 80 foot slippery, grassy cliff and stacking them into my old camper van. I saved until the last the black leather case containing the diaries and notes I had amassed for *Last Wild Years,* the book I was planning to write about how my Scottish odyssey came to an end. As I placed it on the seat beside me I knew that soon, in a new place far away, I would have to start writing it, and making clear my thoughts on how Scottish conservation *ought* to be tackled.

It was too late in the day to start the long drive south. Instead, overcome by a desire to try and see the whole saga in perspective before I finally departed Scotland, I decided to detour 26 miles to the west and camp for the night by the little sea pier that overlooked the beautiful island of Eilean Shona, where it had all begun almost twenty years earlier.

Before that I had been living the fast globe-trotting life of a successful Fleet Street journalist, mixing life, drinks and copy with the elegant, the swift, the rich and the most famous of the day – Sophia Loren, Ava Gardner, Brigitte Bardot, Doris Day, John Wayne, Robert Mitchum, Steve McQueen and Elvis Presley among the celebrities I had known well. (See my autobiography *My Wicked First Life*.) I had lived for periods in Rome, Paris, St-Tropez, Vienna, Madrid, New York, Las Vegas and Hollywood. So long as I secured 'The Story', the world had been my oyster. Then, disillusioned with the high life, morose after a broken romance, I longed to return to the realities of the wild and my boyhood love of nature.

First, I went to Canada where, from the clifftop log cabin I built above the shores of the Pacific, I trekked up wild inlets after grizzly bear, cougar, caribou and bald eagle. It was a preparation. To live such a lone life in the wilds somewhere in Britain was always the special challenge.

I was 42 when I first came to this little stone pier and looked out across the strip of emerald sea to the wooded island, with the golden October sun setting behind the Cuillin hills of Skye. I was a wilderness innocent in those days, but paradise it had seemed to me then, and paradise it looked now. In between lies a story I would never in my wildest dreams have imagined to be possible when I began the odyssey I have recounted in twelve books.

In my four years on Shona and thirteen years at Wildernesse I climbed over 24,000 miles of steep mountainside with heavy camera packs, spent more than 1,000 hours in precarious hides overlooking 42 eagle eyries. I shared Wildernesse with four wounded foxes, three injured owls, two kestrels, a wildcat and a badger. Little realising it would be only the first of 10 wildlife movies I would make over the years, I exposed 14,000 feet of 16mm movies of the eagles, the pine marten family that visited me daily, the rare black-throated divers that were driven from their main nest by fish farmers, the buzzards, otters, wildcats, seals, red deer and many other creatures. By watching in the field,

I discovered many things that contradicted the prevailing notions of science and passed on my findings to a self-satisfied fraternity of so-called experts who were rarely welcoming. And now, in Scotland, it seemed all over.

As the golden sun began to sink and wink behind the great pines of Shona island I had no inkling whatever of the extraordinary wildlife years that still awaited me – that I would spend the next five years in wild mountains all over Spain tracking brown bear, wolf, wild boar, lynx and rare eagles for a book and two movies; or that the following year I would spend the summer filming nesting peregrines in Cornwall. I could not know that I *would* return to Scotland and spend the next three years living in a remote cottage in a Borders forest six miles from the nearest public road. There I studied nesting goshawks and bred barn owls, managing to establish three successful pairs in the wilds where they had long been absent. I had homes, or spent long periods, filming wildlife in Spey bay, the Poolewe-Ullapool area of the north west Highlands, the sea coast of Dumfries and Galloway, and even made a movie of the rarest wildlife in deepest Sussex!

Every year no matter where I was, I would return to my old area of the Highlands to study and film my favourite golden eagles, so notching up (if I include Spain's eagles and vultures) over 3000 hours in eagle hides. When I celebrated my seventy-sixth birthday watching and filming the rare white-tailed sea eagles on the Isle of Mull, I felt I should go out on a high and retire from filming wildlife. Shortly after, to make sure, I sold all my camera gear.

It seemed appropriate, given sufficient time, that I tackled the memoirs that scores of readers over the years had asked me to write. I am still studying wildlife of course and feel I can be excused for some feeling of pride when I see 27 different editions of my books in the bookcase accompanied by 10 wildlife videos. They are a reassuring reminder of the wilderness saga that dominated the last half of my life.

How it all began is told here in *Between Earth and Paradise*. For years the book has been out of print, and in that time I have received more letters asking for it to be made available again than on almost any other subject. I have trimmed it a little to lose some anachronisms and straighten out a few ambiguities, but essentially it is the story as I originally recounted it – of the halcyon years when all was fresh and new and I learned quickly what it really takes to be truly free.

Mike Tomkies

1

Homecoming

The gales were blasting over the island from the north-west, whipping off the tops of the Atlantic billows, and from the mouth of the sea loch, three miles away, deep trough-like waves were roaring up to where I stood, their crests foaming and frothing, until they smashed themselves with elemental fury at the rocks below. It was as if the ocean had declared war on its ancient rival the earth. From the low black clouds scudding across a sky shot with grey and violet, intermittent squalls of hail were being unleashed, the stones hitting the old stone pier and my boat with such force that some of them bounced three feet back into the air. The afternoon passed slowly. I noticed that after each hailstorm the wind lessened for a brief while. If I was to get across at all that day I would have to risk it during one of these slight lulls.

I had been waiting for twenty-two hours. The island's big boat had almost gone down in the previous day's storms and I had spent the night in my old Land Rover, parked above the pier.

It was now 5 p.m. and would soon be dark. The tide had been on the ebb turn for two hours and if I left it much longer I wouldn't be able to clear the sandbank to the west. At low tide my boat, which had dragged its rear anchor in the gales, would be on the rocks below the pier. The next high tide would occur in the dead of night, the pier would then be covered again and the boat would be at the mercy of the rough seas. So I now had little choice but to go.

After the next hailstorm had hurled down its worst and a clearer patch was heading over, I hauled the boat into the pier, loaded in my backpack, sacks of supplies and the fuel tank, hauled up the anchor and set off. Even going slowly, head-on into the foaming waves, the boat banged up and down in the troughs so violently it was hard to keep my seat. By the time I was above the sandbank another squall hurtled over the mountains. Now that I was moving against the wind the hailstones were hitting me hard, as if thrown down by a vast angry hand. I had to shut my eyes, for it was like being stoned by hundreds of small pebbles.

With one hand on the tiller, and squinting through the barely opened fingers of the other held across my face, I thought I would perish in the roaring noise for I could hardly tell the difference between the hail and the lashing spray, the stormy grey of the air and the green of the icy water. For a time I had to steer by the feel of the waves and the wind. I limped into the lee of the island, turned west for the last two miles, negotiated the narrow channel between two rocky islets where resting seals were reluctant to hump their way into the stormy water and, after another brief blasting from the gales blowing down a valley, finally reached the shelter of my little bay.

As I hauled my boat into the special cradle of planks and rope I had made to protect it from the rocky shore, knowing that yet again I would have to return at around 2 a.m. and haul it above high-tide level, I fell a surge of gratitude. In this wild remote place, just to reach home safely with some supplies was merciful deliverance.

With loaded pack and hand-held provisions I puffed up the steep 300-yard slope to the old wooden cottage, my trousers drenched beneath the cheap storm suit. I smiled ironically – to think I had imagined the Scottish Highlands would be a more benign environment than the Pacific coast cliff-top in Canada where I had spent three-and-a-half exciting years!

Indoors I lit the paraffin lamp. As the room sprang into bright focus and the beam shone through the window upon the ash trees that lined the burn and shielded the croft from the worst of the Atlantic gales, I glanced at my calendar. It was 20 October:

precisely a year since I had first arrived in Scotland in my old Land Rover as part of an odd pilgrimage – to take a second look at the islands of my birth and to meet the author Gavin Maxwell, whose *Ring of Bright Water* had suggested to me a wilderness life in Britain was really possible. A year ago to this very day I had descended the bleak lonely hill above his former home, Camusfearna, and found the cottage gone.

I lit the fire in the open hearth and sat back staring at the flickering flames, reflecting what an extraordinary year it had been for me. It seemed like a decade since I had watched the sunset from the drift log cabin I had built on the British Columbian coast, wondering what to do with the rest of my life. Dissatisfied with my hectic life as a Fleet Street journalist in the hedonistic showbiz capitals of Europe and in Hollywood, I had gone to Canada four summers earlier seeking tranquillity in new surroundings to write a novel. But that is another story . . .

Although the book failed, during the years of working on it my youthful love of nature had been re-born, and I had known some thrilling experiences with wildlife, culminating in treks into one of the last true wildernesses in all North America. I had watched grizzlies, bald eagles, cougar and caribou in their remote fastnesses, both alone and in the company of an old Scots-Indian called Tihoni, who taught me more by example than word. To me, those treks could not be repeated, and when they were over a strange sense of anti-climax had pervaded the months that followed. I had learned to live with and to love the wilderness, but the time had come to move on.

One morning I had woken to a faint banging sound. Through the window I saw a bulldozer clearing trees on the once-deserted northern spit of land over which the bald eagles flew to their nest. A speculator had bought the area and was building vacation chalets. That afternoon on the little island where I collected my small ration of oysters, I found a new notice board: 'Private Oyster Lease. No Trespassers'. There had been unprecedented queues a mile long at the ferry crossing from Vancouver that weekend. More city-bound folk had discovered the beauty of the coast and it was fast becoming a major vacation area.

With memories flooding back of my boyhood in the woods and fields of Sussex, I decided to sell my land lease. Although replies flooded in from folks who thought they wanted to leave city life, asking questions like – what shape is the cabin? where can we park the car? what cooking fuels are available? what furnishings are there? – the few that actually came to see it were portly middle-aged couples who had spent time in the beer parlours. Blowing like whales along the trail, talking of their youthful fishing in the wilds, they would take one look at my unstately home, at the rugged terrain, the lack of mains water, gas, electricity, phone and road, and that was that. Most of them had no intention of buying; they were bored and just wanted somewhere to go while up on the coast.

I had almost given up hope of selling when one afternoon I saw children stumbling along my beach, followed by a couple and a dog. The man was an English professor at a Vancouver university and he told me they had been looking at a piece of land similar to mine but with no beach or cabin. They were being asked $3,500 for it.

'You can have this land, cabin, furniture, boat, slipway and this beach for a thousand less,' I said.

The professor was puzzled. 'You mean a thousand more?'

'No. I mean a thousand *less*.'

He came up the log staircase and wrote out a cheque there and then.

I felt a strange mixture of emotions on arriving back in Britain – the tiny roads, squashed soggy little houses of London's outskirts, the little cars, delayed luggage and frowning airport faces – so many people crushed together. But one thing struck me as never before: despite her economic troubles, Britain had preserved an extraordinary outward-looking attitude. One could travel all over North America and never hear more than a few words about Britain on radio or TV, or read more than a paragraph or two in the papers, yet in Britain most of the news was about other countries. The first radio programme I heard was 'Any Answers' – ordinary enough, but hearing the well-balanced and quietly expressed views on a variety of topics from people mainly in small towns and villages throughout the land, not the overheated big cities, made me feel Britain's heart was beating as soundly as ever. A big heart that would yet absorb the extremes of Left and Right and the excesses of the modern age, and still beat stoutly for individual freedom. Like many a returning emigrant, I was feeling intensely patriotic.

Adopting the attitude of the Canadian logger after long spells in the wilds, I had promised myself three luxury days in a posh hotel with a heated swimming pool before I looked for cheaper accommodation. But I never had them. I visited an old friend, photographer and entrepreneur Alex Sterling, an extraordinary character who had toured Mexico and Spain with Pietro Annigoni to produce glossy books that contrasted the arts of painter and photographer, and was a friend of the Maharajah and Maharanee of Cooch Behar in India. Alex once talked showman Mike Todd into making a film using Elizabeth Taylor as a model! He also ran an exclusive supper club in South Kensington for theatrical stars and as we talked long into the evening, recapturing our experiences among the fast-living set I had abandoned for the wilderness years in Canada, I told Alex I would soon be looking for a cheap London base. He seized my arm and propelled me through a door into a thickly carpeted room, empty save for racks of wine bottles and a camp bed.

'It's yours,' he said. 'Free, for as long as you like. But tell me if you drink any of the wine so I can change the inventory.' What an unexpected windfall! Next day I bought a small camp cooker, an axe, tools, sleeping bag and pots and pans for my safari into the 'wilds' of Britain. Then I purchased a secondhand Land Rover, took out the rear seats and built in cupboards, shelves, a clothes cabinet and a folding plywood bed.

In less than a week I was in the Lake District and after much searching, hired a caravan on the banks of Esthwaite Water, a small but serene lake. That first day I tried fishing but had no luck. Sitting there in a little polystyrene boat, the low soft-green hills of England around me, I suddenly felt strange, oddly dissociated from the real world, like a being from another planet. Where is 'home' anyway, I wondered.

Several weeks were spent exploring for an isolated waterfront home, driving and camping rough round the lakes. Finally, after two months and some odd encounters, I decided the Lake District was too highly organized, too crowded and far too expensive for my slender resources. I would go to Scotland, continue my search there and meet Gavin Maxwell. Then, on an October evening in a Lake District pub, one of two young nurses started talking to me about my plans for going on to Scotland. She then asked if I wanted to visit Camusfearna, which seemed more than coincidence. I said I had planned precisely that, and that I also hoped to meet Maxwell himself, to see if we might partake in some worthwhile wilderness venture together, for I felt he had made many unnecessary mistakes.

'You know he's dead?'

Her question shocked me. I had read that his home had burnt down and that Maxwell had moved to a lighthouse but I knew nothing of his death from cancer in hospital at Inverness in the very week I'd made my decision to return to Britain.

I drove up through Cumbria and across Scotland on a dingy, grey day. Parking by a Forestry Commission fence, I climbed over and pushed my way through the thick, young conifers on the hill behind where I thought Camusfearna lay. After half a mile I saw a telephone line. By following it I came to Maxwell's overgrown track through the trees. I was surprised that he had had a phone.

Camusfearna looked dark, lonely and bleak. The cottage had been razed to the ground, One rough stone lay over Maxwell's ashes, and a wreath of fading green leaves and a wire cross with conifer leaves interwoven in it were the only signs of the man who had created a concept of paradise for so many. Nearby, its surface covered with the scarlet berries of the rowan tree on which Maxwell believed he had been cursed, stood the stone monument to Edal, his famous otter who had replaced Mijbil. And engraved on it Maxwell's inscription silently exhorted the dark sky and the rowan tree: 'Whatever joy she gave to you, give back to Nature'.

At that moment I felt an impulse to try and breathe new life into the place, fill it with love, resurrect it. I saw another old croft on the site which Maxwell had not mentioned in his book, and it seemed right for me to try and acquire it, so that I could restore Camusfearna to a new beauty. Not for many years, until I had read Maxwell's later books and the biography by his friend Richard Frere, did I realise that superb lyrical writer though he was at his best, I would have little in common with Maxwell as a wild-animal keeper, a wilderness dweller, or as a man. But I did not know that day. I paid a visit to the titled owner of the land, who seemed amused at my direct approach, and thus began the fruitless negotiations which ran into weeks.

Back in London I met again the girl I had loved before emigrating to Canada, who was still unattached. Then a leading popular magazine offered me a writing contract for six months. Flattered at being remembered after so long an absence, needing finance for my wilderness life in Britain, I accepted and moved into a flat in Knightsbridge – for that was what the lady wanted. But the renewed romance fared little better than before, and taking the job proved a mistake. I had been a free wilderness man for too long. One can never go home again – not on the same terms. I felt like a man in a cage, sold the flat lease again, lucky to escape after only three months.

I made contact with Frere, now one of Maxwell's executors, who had a small cottage for sale on the author's estate. We met at a lonely crossroads near Glenelg, but the cottage was not what I wanted. It was half a mile from the beach and one of the ferry roads to Skye ran between it and the shore. It could never be the small wilderness paradise I sought to re-create.

I travelled the Highlands, living rough, finding places with names where there were no longer the places, feeling a growing empathy with the landscape itself, its misty, mystic beauty, its hidden wildlife. I just missed a lonely seaside cottage which had stood empty for six years on the Inverness-shire coast. An aged couple said they might sell an old house four-and-a-half miles from the road on a sea peninsula when they were too old to reach it in the summers. They were in their 80s! After one night's camp by an ancient croft at Port Luinge I woke to an almost identical view to my place in Canada, but I had no luck with the absentee landlord of the 40,000-acre estate on which it stood.

I began to feel I was on a hopeless chase, but decided to try once more before returning to Canada. In March I visited the Forestry Commission office in Inverness where I was shown a huge bundle of letters, all similar enquiries to mine, though this time I was given details of an isolated forestry cottage by a fresh-water loch in north Argyll, which was to be put out to public tender.

I found myself heading south to meet an estate manager who lived at Dorlin on Loch Moidart. As I drove round a bend the road suddenly opened out on to a view so magnificent, right down the sea loch and over the islands of Eigg and Rhum, that its beauty hit me like a stroke in the soul. Entranced, I stopped the Land Rover and impulsively knelt down to kiss the ground, to touch the rocks on the shore. A strange excitement grew within me for I felt in some odd way that I was actually coming home. (Although I knew I had some Scots blood, I didn't know until a talk with my father over a year later that my mother, Adele MacKinlay Stewart, who died when I was four, was pure Highland, and that her father, John Stewart, was born on Islay).

When I met the estate manager it seemed I was doomed for more disappointment. Phil Corcoran told me his estate had nothing to offer me. I turned away and looked out to sea. A shimmering blue vista of myriad islets in the loch led out to the great Atlantic. To the north lay a three-mile-long island, partly covered with many different kinds of conifers. For anyone who loved trees and the sea as I did it was clearly a paradise. The ruined shell of ancient Castle Tioram, stronghold of the Clanranald chiefs, stood silent sentinel over an idyllic scene.

The lovely island was called Eilean Shona. I was told that its owner, Digby Vane, was a retired economist and a one-time Olympic reserve sprinter. I prevailed on Mr Corcoran to telephone Mr Vane and the upshot was that he agreed to meet me the next day.

In blazing sunshine and a stiff westerly breeze Digby Vane roared across the mile of sea from his home on the south-east edge of the island in a triple-hulled dory and pulled up with a grin. I think we liked each other on sight, and he invited me back to look and meet the family. We bounced over the seas at a terrifying speed and I had to cling with both hands under the thwart to stay in my seat.

I met his wife Kay, and their partner, inventor and pioneer in radar monitoring systems, Reggie Rotheroe, and went to talk in their antique-filled house. The Vanes had bought Eilean Shona (the Gaelic Eilean Sean-ath means Island of the Old Ford) from Lady Howard de Walden in 1962. As we talked, Kay told me that *Peter Pan*'s author, J. M. Barrie, had written the first film script of the book and part of his *Marie Rose* in that very drawing room. They still had Barrie's desk. 'He called it the most beautiful island in the world,' Kay added. If it was good enough for Barrie, it would be more than good enough for a writer like me, I thought.

Later, as we walked in the rhododendron-filled garden, by the lily pond flanked by large and rare meta sequoia and Japanese cedar trees in which otters sported at dawn and dusk, Digby Vane told me there were seventy different types of conifer on the island. Many of them had been planted personally by a previous owner, seafarer Captain Swinburne, who had collected some of them on his China trading runs. It was reputed to be one of the most comprehensive collections of conifers in Britain.

We skimmed away in the dory along the island's superb coastline, then staggered over seaweed-covered rocks to look at an old cottage which nestled in a green dell. It had no doors or windows, the skylights were broken in and sheep were using it as a shelter.

None of this bothered me but I said it wouldn't do because when I turned round I could see a single-track road and a couple of cottages on the mainland.

On the Atlantic edge of the island, two-and-a-half miles from the little mainland pier, we came to a secluded natural bay with high rocky walls on either side. Almost three hundred yards above the sandy beach stood a large dark wooden cottage – broken windows, rotten boards hanging from rusty nails, and a corrugated-iron roof. We plugged our way up the steep bracken-covered hill to it and I turned to look at the view. I could see over the whole mouth of the sea loch, a vast 180-degree panorama with not a house, road or sign of human habitation in sight. From a hill to the west I could look out over the Atlantic and, again, the mist-shrouded islands of Muck, Eigg and Rhum.

I stood there spellbound, feeling that pang, almost of pain, of first love. A row of tall ash trees protected the cottage from the western blasts without spoiling the view. An agate-coloured burn gurgled past the cottage, provider of the most valuable element for a wilderness living. Overhead a large, speckled buzzard soared, it too following its own lonely star.

'There's nothing between you and Canada here!' laughed Digby, regarding me quizzically. 'You're not a *smuggler*, are you?' I laughed too, and assured him I was no smuggler. Already Canada was a fading dream.

'What's this place called?' I asked him.

'Ballindona,' he answered.

'Ballindona?' I repeated. 'Like Bella Donna. Beautiful woman. Well, I don't have the beautiful woman but this place would be paradise enough for me!'

Digby coughed quietly. 'Well, you seem a reasonable sort of chap. If you're crazy enough to do the place up and live alone all the way out here, you're welcome to it.'

We looked over the cottage. Much of its woodwork was rotten. There was no road, no electricity, gas, phone, bathroom, kitchen, sanitation or even piped water. Sheep dung five inches thick covered the floor. A dead sheep's carcass propped up the shattered door. As I picked up the skull by a horn the hoofs rattled down from the tocsin skin, making my first entrance macabre. It was going to be hard work; all the building materials, timber, tools and personal possessions would have to be ferried over two-and-a-half miles of uncertain seas, then carried up the steep 300-yard slope. But nothing could have dissuaded me.

Over lunch at the big house the arrangement was made. I could have it for at least three years, unless they had to sell the island for economic reasons. I need pay no rent, just repair the place well enough to live in myself.

I had only £297 left to start my new life in the Scottish wilds. I lined up some journalism in London, then hurried back to the Highlands. The Land Rover was loaded with polystyrene boat, window glass, axes, saws, pickaxe, sledgehammer, plastic loo (shipped from Canada), water carriers, clothes, books, bedding, fishing gear, paraffin heater (found on a London dump and the best I ever had), oars, gas canisters, tinned food and a portable cold box. At a saw mill near Fort William I strapped on three dozen heavy timbers and slowly groaned the last forty miles of single-track road to the little pier opposite the island.

Next day, regarding my tiny boat dancing like a feather on the choppy ocean, I hired the island's two workers and the big 30-foot boat to take the heaviest gear and timber to Ballindona. As the two men conversed volubly in Gaelic, I felt more 'foreign' than when I had first landed in Canada! Even at high tide we could not get in to the beach, they said, so everything had to be left on a high hump-backed rock overnight.

In the morning I rowed the two-and-a-half miles to the croft
in strong westerly winds. The little boat might have been adequate
for the Lake District waters but on that wild sea I was dicing with
death. By the time I arrived I felt ready for bed but had to spend
the day hauling up the timber, glass, cement, hardboard and
other gear in case the tidal waves smashed the lot along the rocky
beaches. Thirty-eight times up and down the steep slope, which
in places was one in two, added up to over 30,000 steep feet, and
as half of it consisted of perspiring upward movement with
heavy loads, I reckon I became a semi-fledged Scots hill-walker
on that first day. Once, carrying sheets of hardboard on my back,
a gust of wind sent me staggering across the burn. So I made a
sled of two timbers and hauled them up four at a time, using a
rope harness. My heart pounded, my legs gave out twice, and I
knew the meaning of Scotch mist before the eyes.

During the week it took to make one room habitable I slept in
an old caravan on the shore below the big house. With a window
missing, it was cold and I returned to it soaked with sweat and
rain. With nowhere to dry clothes, I donned new gear the next
morning and the following day had to put on the half-dry clothing from the previous day. One night I woke with the caravan
being tossed around like a ship at sea. The island's six rams were
using it as a rubbing post. On the third morning I borrowed a
small, 1½-hp outboard motor, to help when rowing against the
wind, but the beach presented problems. After clearing a channel
through rocks with a crowbar, I found later that lots of small
rocks had drifted in again on their seaweed umbrellas, necessitat-
ing further clearing.

The fourth day was so stormy I had to walk in over the mountain trail, a sheep and deer path through quagmire, peat bog,
spiky gneiss rocks, and over slippery stone causeways with two-
foot gaps between many of the steps, which gave problems to a
man with a full pack.

Shovelling out sheep dung, removing the window frame and
re-glazing and painting it, fitting hardboard round the walls and

rebuilding some rotten foundation sills, I finally had one room in fair shape. One afternoon I nearly lost the little boat when I underestimated the speed of the incoming tide. I just managed to reach it with a long pole, standing waist deep in the cold sea as it was about to depart for the open Atlantic.

But on the lonely walks over the hills I was continually rewarded – by the sight of three red stags moving so effortlessly over the rugged terrain that they seemed to float as if obeying the strings of an invisible puppeteer, by a kestrel hovering in the wind like a tiny anchor in the sky as it scanned for mice or frogs below, and by the buzzard soaring over my head.

On the eighth day Reggie Rotheroe came down to the caravan with a little radio he had made himself. 'You'll be needing that up there,' he said, plonking it down and leaving almost before I could thank him.

The Vanes allowed me to borrow some items from a boatshed of furniture near the Shona pier. I picked an old iron bedframe, a mattress, two rush mats, a kitchen chair and a superb mahogany four-drawer desk, the very one on which James Barrie had worked. With the help of their chief man, Iain MacLellan, and the big boat, we shipped them over and lugged them up the hill. So I moved into my new wilderness home.

Even after the years in the Canadian wilds, the lonely wooden cottage seemed a ghostly isolated place that first night. A hundred years earlier a small settlement of crofters had lived here and the ruined stone walls of three of their primitive homes still showed above the bracken and rushes. I woke up after a few hours, scared because I was sure I could hear the distant skirling of phantom bagpipes. But the noise was caused by the wind whistling through a knot hole on the window frame. Even so, now that I was some two miles from the nearest people, the isolation seemed enormous.

Next day the rain and storms had gone. I went out and stood on the green hill overlooking an ocean now beautiful and still. The early morning sun from the east dazzled the myriad islets and upon the cerulean blue of the sea loch. Below me, down the path that followed the singing, winding burn, the waves broke gently emerald and crystalline over the white sand of the beach that now appeared to be lit from within, as if by some heavenly radiance. The feeling was strong upon me that I had indeed come home.

2

Getting to Know You

A May shower had just passed over and the sweet smell of the earth, after the benediction of the rain, mingled with the tang of the sea air, as I set out in goldpan sunlight for a walk to the Atlantic edge of the island. As I trod the gleaming deer path I heard the waves chuckling on the rocks below and lapping the small gravel bays, the tiny burns tinkling as they etched their way through ancient granite to their parent sea, the wind breathing through the heather. These sounds were my only music – but a kind one could never hear in a concert hall.

How different were the stunted trees and these open hills, still bare with winter grey, from the mighty western Canadian forests, the plunging torrents in the rivers when the winter snows began to melt, from the long tidal flats where the great grizzlies would now be grazing on the first grass at dawn. Although I had studied English wildlife in boyhood, I knew little about the Scottish Highlands. I felt filled with a sense of discovery and open to new influences on my first exploratory walk on the island. What wildlife would I find here?

Suddenly I heard a strident voice ahead. I peered through the heather. There, in a natural grotto of grey lichened rocks, perched on a fern, was a tiny wren singing in stentorian tones far too loud for its size. With the rocks reflecting his voice like an amplifier, he sounded like an imitation canary, his whole body quivering with the effort of producing vibrant song. Another wren, presumably a female he was trying to impress, for she was silent, flitted to a fern nearby. He flew to a little domed nest below some heather on a rock and, spreading his short tail feathers for support, flirted his wings in the sunshine, as if enticing her to take a look. He succeeded for she flew straight in to the nest, but a few seconds later came out again and zipped back to her perch. He immediately flew towards her then veered off, whirring like a big, brown bee, towards a small waterfall into which he seemed to vanish. She followed but in less than a minute both flew out again and disappeared into some bushes in fast dodging flight. I went over to take a look.

The wren had built a nest beside the waterfall, so close that some drops of spray were falling on to its roof. Inside it was totally dry. No wonder early naturalists believed wrens used their saliva as gum to build waterproof domes on their nests.

Wrens, like their namesake Sir Christopher Wren, architect of St Paul's Cathedral, are master builders but only the males do the nest-making. Women's lib is a relatively new concept in human life but in the wren world it has been the *status quo* for thousands of years. Perhaps they imported it from America for the wren is one of the few European birds to invade the Old World from the New, crossing the then existing land bridge at the Bering Straits in the Pleistocene Era. At any rate, the little males build several intricate domed nests then have the task of enticing a female to use one and be a wife. So good an engineer is the male that he will gather wood fibres caught round fallen trees in rivers because when they dry out they bind the nest more strongly. A really vigorous male in his prime, in areas with plenty of sites and food, practises polygamy and can support several wives. But, I was to find out later, this almost never happens in the rugged Highlands of Scotland – just as the human divorce rate and extramarital adventures increase in affluent or crowded city areas, in places where a living is harder to win, wrens have as much as they can do to support one wife and a large family.

As I walked on, the first humble-bee queen of the year, sluggish from hibernation, zoomed past, looking for a suitable site, like an old shrew hole, to make her nest and start her new colony.

Suddenly I heard a strange sirening noise, a loud drawn-out '*Awhoo*', and at first I thought it came from the seals that hauled themselves out on the rocky islets at the loch's mouth. Surely only the grey seal makes such sounds? There was a commotion on the placid surface of the sea below – a splendid black-and-white eider duck drake was flap-chasing a dowdier brown female, his thick bandy legs thrashing the water as he seemed to be trying to grab her tail in aggressive courtship. The duck just swooshed into the water again and the drake went down too. Through my field glass I saw him swim urgently around her, like a hefty little tugboat, tossing his massive wedge-shaped beak skywards, as if he were gulping. He really was a solid barge of a duck, decked out in superb evening dress, his black jacket a startling contrast to his white back and snowy velvet cheeks beneath his black-capped head. On his white breast was a patch of delicate peach, and overlaying each wing were long white plumes, fine enough to grace the hat of the Laughing Cavalier. Occasionally he reared up, showing the female his black velvet belly, and flapped his wings. The lady, however, seemed unimpressed for she paddled stolidly along, her bright eyes shining like buttons. As I continued westwards into ever brighter light from the broad Atlantic, his gentle '*awhoos*' came like a soothing serenade from the sea.

Now three buzzards were circling overhead, lit bronze from the reflected light – a large female attended by two smaller birds which I took to be males courting her, for birds of prey finish training their youngsters before winter is through. The two males showed no animosity towards each other, all three flying along together. If she was not already paired with one of them, she would choose soon.

I crossed two small ridges and found myself looking over a sublime natural lagoon. The tranquil, crystal water was the colour of jade near the rocky edges, deep as the blue of heaven out in the centre and light amethyst where it rippled gently on to a beach of white-shell sand. The whole floor of the shoe-shaped lagoon was also of this crisp sand, with just here and there a clump of green or brown tangleweed growing up like inverted chandeliers from the sea floor. It was now high tide and nowhere was this 80-yard pool more than six feet deep. The little beach itself was protected on all sides by high rocks with a patch of green sward on its rear edge, like a preliminary carpet before you reached the sand. No South Sea island could boast anything better than this, I thought. It was as if the retreating glaciers had wanted to show they could fashion a bijou beauty too.

I spotted something in the water, the broad head of an animal swimming, making an unusually wide V as it went. Not since boyhood on the river Adur in Sussex had I seen a British otter swimming, but the huge V seemed too large. Through the field glass I saw it was indeed an otter but the illusion was caused by the otter having a little object, probably a butterfish, in its mouth while it swam along the surface. As it climbed out of the water, wet hair clumps dripping, and began to eat, another otter a third of its size came up curiously. The mother otter stopped chewing, as if to let the youngster have a piece itself, but apart from sniffing gingerly it showed no desire to eat. When the mother finished her meal, she ran up and down a rock ledge less than a foot above the water. The youngster followed her a few steps each way as she went past each time. The mother seemed to be trying to nudge the baby into the water – but it didn't want to go. The adult then closed her forepaws against her sides and slid off the ledge. She dived, twisted, turned and rolled in the water with joyful abandon, as only an otter can, perhaps showing the youngster what a wonderful watery world there was to play in. Her bairn showed no wish to follow. Out came the mother again and once more nudged her reluctant child. It seemed odd there was only one, for otters usually have two or three. It dug its little web-footed claws into the ledge for all it was worth. I wondered how long it would take for the mother to coax her baby into the sea. Suddenly she turned, squeaked so shrilly it was almost a whistle, and both she and the youngster went lolloping across the rocks and out of sight on the other side. I hadn't moved, the slight breeze had been from them to me, yet somehow she had sensed my presence.

As I passed the wrens' nests on the way back the little male came darting fussily from one heather spray to another beside me, his wings making noisy '*brrrt brrrt brrrrt*' sounds. For all their 3 ¾-inch size, male wrens are fiercely territorial and in the breeding season will frequently see you off, like a little suburban dog escorting the postman off the premises.

I cut down to the shore and after a few hundred yards came to a lovely secluded place, sheltered on both sides by high ridges covered with heather and rows of stunted trees. To my surprise there were six different kinds – oak, hazel, ash, rowan, birch and, nearest the shore, alder. None of them topped more than 20 feet yet they were old. They were also the last trees growing westwards on the island and seemed to have great characters of their own. Each had found a precarious foothold on the thin soil or, blown as seed into the rock crevices, had slowly thrust roots down every fissure to find a mite of sustenance.

They were dwarfed from surviving a harsh environment, bent from battling the prevailing south-westerly gales and crooked because every inch had adapted to slight variations in the passing seasons. Each tree was unique in its own way; the alder short and thick-trunked, the hazel bushes tight and canny, and the rowan like a witch's broom – they were small, but none had been defeated. The alder buds were the first to break and show the green of future leaves, the ash and oak still as tight as winter, I felt I could love them even more than the great forest trees of western Canada, growing tall and uniformly straight amid such protected abundance. These tough little characters, as honed as flyweights, had put up a brave fight to live, had earned the right to be there.

As I continued on my way home I was conscious that I had not yet earned that right. I didn't fool myself into thinking that because I was prepared to live alone on the remote edge of a Scottish island, two miles from the nearest humans, and to cross

the stormy loch in a small boat, these things in themselves gave
me any rights.

To me, living in the wilderness was a privilege. I felt these wild
Highlands would help me only if I lived in them with reverence
for their beauty. They might help me, as had Canada, to find a
way of life in which I could avoid disturbance, temptations and
the social structurings of the city and be inspired, and gain time,
to think and to write. Time was paramount for I was well into
my forties. Once I had the old wooden croft in shape, I could get
by on £7 a week, though I still had to earn *that*.

Bouts of work in a Sussex animal market, in a saw mill, as
cook on a yacht, and in Canada as logger, salmon boat deckhand
and assistant blaster, had all been good experiences but had
shown me that nothing inspirational comes from low-paid labour
done merely in order to survive. I briefly considered the usual
Highland jobs like postman, road or forestry worker – and writing at night. But I had run out of youthful energy and anyway
living on an island made such jobs impractical.

Feeling an interloper, I reached the green hill of the croft again,
where the little banks of wild daffodils, some broken off by recent
gales, were all nodding in the breeze as if to greet me – the only
gentle reminder that once a family had lived here. Not for 33
years had they heard the sounds of children's tears or laughter,
the tired tread of father, heavy from work after his walk across
the dark island with an oil lantern in his hand. Now I had come,
no part of the old crofter life, and yet the lonely flowers made me
welcome. I watched their yellow heads nod away and towards
each other, as if they were whispering in the winds that stirred
their green leaf skirts. I knew, no matter what men had been
doing in the busy cities miles away over the sea, this brave little
band of flowers had burst forth each spring, hoping perhaps that
someone, some time, would come along again and be inspired by
their own struggle when all had often seemed lost.

There had been no daffodils on the lonely cliff in Canada but
these reminded me of the daffodils of my own country boyhood

in Sussex where my love of nature began. Suddenly they seemed an emotional link with those happy carefree days, and I felt oddly grateful to them. Then I smiled at a memory. Months earlier I had taken the girl I loved down to see my old nature haunts. 'You'll never be happy unless you get back to that life for good,' she had said. But an isolated life in the wild Highlands had no appeal whatever for her. One evening I had taken her to see the film *Paint Your Wagon* and when grizzled old timer Lee Marvin had grunted out 'Wanderin' Star', with lines like 'Do I know where hell is – hell is in hello. Heaven is goodbye for ever, it's time for me to go', she had nudged me. 'That's you!' Lee Marvin! Unfair, I thought, for I *had* asked her to marry me. Yet she had a point – before the Canadian years I travelled and worked in 18 countries.

I chuckled as I walked into the lonely cottage. Well, I'd lost the lass for she was now married to someone else. If I was now to live alone for good, what an idyllic place in which to suffer.

Yet to earn even a small living I was still trapped in the trade for which I was known by publications in nine countries – writing about the movie world and the famous. To me, a great boring porridge of kitchen-sink concepts, the elevation of the moronic life into drama and the cult of the anti-hero, had invaded much of literature and entertainment in recent years. It had taken me a long time to break away from all that, and I didn't want to write for a living about the new actors who had emerged, excellent craftsmen maybe but who were often so boring off screen. The era of colourful characters such as Wayne, Gable, Mitchum, Bogart, Ava Gardner, Marilyn Monroe had passed, and prudent technicians had taken over. I had grown older too and it wasn't fun any more. The once 'glamorous' theatre world now appeared to me to be only a few acres of grey buildings in various capitals where some folk were overpraised for pretending to be someone else; and so few films or plays had any real vision.

All the same, for more than a year I still had to dash down to London, where I retained a small bed-sitter. I raced round studios, sold my articles, before hurrying back to the island and what to me was the last reality – the natural world. I dislike grants or subsidies, especially for writers or artists. I feel if you can't make your own way you don't deserve to do such work. Therefore, I had to overcome my new dislike of the work to which I had once been attracted. But the visits to London grew shorter and each time I returned to immerse myself slowly into Scottish wilderness life.

Slowly was the operative word for while the Canadian years had equipped me better than if I had come straight from English city life in the first place, the animal life of these open hills and small forests was quite different. The contrast between overall 'harmony' of nature and its 'callousness' towards the individual life, was repeated here but on a more finely-etched scale that was somehow more personal.

One March morning, as missel-thrushes, robins and chaffinches filled the little glen with song, and the sheep around the croft, soon to have their first lambs, began to bleat more noisily, a racing pigeon with red, blue and green rings on its legs flew in, fluttered down weakly and began pecking at scraps. As I went down to the shore to fetch a gas container, a small tortoiseshell butterfly that had wintered in the cottage's roof flew out beside me into the spring sunshine. The racing pigeon flew a few yards, then sat in a cranny between two rocks with its feathers fluffed out. It had flown a long way and was exhausted.

The first bracken shoots were coming through now, doubled over like snakes' heads, as if afraid to show their faces to winds and rain until they had grown stronger. I had been told the area would be engulfed by this weed in summer, its underground roots poisoning the earth so that little else would grow, so each time I went down the slope I kicked the snake heads off. Later, when I went down to re-set the boat on the high night tide, I found the racing pigeon had been reduced to a pile of scattered feathers, with just part of the skeleton and one eye left. I doubted that a sparrowhawk, whose larger female is only fifteen inches long, could have killed and eaten it so fast; it was probably a fox. It seemed oddly sinister that this little murder had taken place only forty yards from the croft when I had not seen or been aware of anything at all. Here was the wild law of indifference towards the weak individual. The pigeon could not rest until it had found good shelter first, and it had been too exhausted to do that. There was no rest, even for the hard-working innocent.

On calm days, when I knew I would be at my desk for hours, I sometimes rowed up the loch after dawn for exercise. One morning, while trying to dip the silly little oars of the polystyrene boat silently in and out of the water past the pine and larch forests on Deer Island, I saw a row of nine bluish dots in the bents and grasses of a tussocky meadow. I was surprised as I slid nearer – why on earth were nine herons, all neatly spaced some seven yards apart, sneaking along as if on tiptoe between the little mounds? In their blue-grey outfits, white epaulettes and black-and-white head plumes, they looked like a search party of little traffic cops from some South American state as they quartered the slopes. I pulled in below a small heathery ridge and looked over – just in time to see one heron go berserk. Raising its wings halfway, like an old lady lifting her skirts, it ran gawkily all over the place, spearing its fearsome five-inch beak into the earth.

It missed whatever it was after but then the heron next to it did the same – darting about crazily, digging its bill into the ground. Evidently it caught what the first heron had been chasing, a mouse or frog, for it leaped clumsily into the air and flapped, long neck still extended, to the banks of a small burn. There it dipped the mouse or frog into the water, elongated its neck straight upwards and swallowed the creature. I could see its neck feathers stick out then smooth down again as the lump went down its gullet.

I must have moved then for the whole line of herons bounded into the air as if of one mind and flew with their strong but deceptively slow-beating flight back to the trees on Deer Island.

Why there? I soon found out when I rowed over and walked below the trees. High up in the pine, larch and firs on the island's easterly shore there was a huge heronry and I counted thirty-eight nests, white droppings splashed below told me that the big birds were now roosting in the nests and the breeding season was starting. Until that day I had always regarded the heron as a lonely figure, patiently standing for hours in water to make one lightning dart with its beak at any luckless fish that came near. In mediaeval Britain it was believed that herons could secrete a special oil from their feet which *attracted* fish! After what I had just seen, it seemed they also occasionally worked as a team. An extraordinary sight, and I resolved to watch them whenever I could.

A week later I rowed back in the pre-dawn twilight with some hazel switches to bend as hoops into the ground and some sacking to put over them to make a 'hide'. I would erect this on a high knoll which was almost on eye level with the top of one large nest, cover it with conifer branches and herbage and so watch the birds while out of sight. I landed on the west edge of the island, away from the heronry and stole through the trees. It seemed the nests were devoid of birds. Believing a few could be sitting close by, for some of the older nests were about five feet across, I kept going slowly. Suddenly, to my right, I saw a large blue-grey gliding form as one of the birds cruised slowly a few yards above sea level, dropped lower, then disappeared. Leaving the hide materials behind, I slid silently forward – to see an extraordinary tableau.

Some flat rocks just off the shore were a riot of colour, ablaze with herons, for 29 of the huge, gawky yet elegant birds were standing on them. This had to be a 'dancing' or gathering ground which I had heard about in boyhood but had never found, the social meeting area which herons sometimes use before the breeding season really gets under way. Each heron stood as if it believed itself to be alone, its long neck hunched back into its shoulders, looking like a disgruntled pelican that had swallowed a rotten fish. Presumably there were both males and females in the party but there was no courting behaviour, no aggression or pecking, nor any sudden running at each other that one sees in gull colonies. Occasionally a new heron glided silently, and as it landed it kept its huge wings aloft and did a clumsy hop, skip and a jump for a few yards before it folded them. Whether this was to help stop its forward impetus or a deliberate dance I wasn't sure. Each standing heron it passed opened its wings halfway, then did a little awkward, long-legged, forward fandango of its own. There was nothing regular about this dancing – maybe they heard a different drummer or one beyond human ken – but it was comical to watch.

Herons are loners most of the year and this odd gathering with its truncated gavottes is said to stimulate their social instincts and their sex hormones before breeding. As I sneaked back to the knoll and quietly erected the hide I felt fascinated by the herons, and each time I watched them over the next few months they seemed more weird and wonderful.

It was about this time a new 'neighbour' came to stay around the croft. Woken early one morning by a loud '*Kaah kaah*!' I looked out of the window to see a hooded crow perched near an old and what I'd thought to be a disused nest of old dry twigs in a birch tree just above my little bay. He kept calling noisily, then suddenly another crow, smaller and more tattered-looking, landed near him. I was sure this was a female, maybe his mate, for crows pair for life, but to my surprise he immediately flew at her, flapping his wings and extending his beak aggressively, and drove her away. A few minutes later she came back again, receiving the same unchivalrous treatment before flying off to the east in apparent disgust. I knew a little about crows, having studied them as a youngster when I'd had a pet crow I called Clark, and also a fairly tame one which had come round my cabin in Canada. When it comes to mating, crows are no casanovas but complete pragmatists and they choose their mates for largely practical reasons. It seemed he had summed up this old female and her abilities to breed, help repair the nest and rear young at a glance. She was old, past her prime, and he didn't want a tired old lady as a wife.

He called loudly during the next two days, and one afternoon I looked out of the small rear room I used as a kitchen to see his raucous clamourings had either attracted a female crow more to his liking, or else she was his normal mate who had returned from elsewhere. While a few sheep stood in apathetic audience, he crouched beside the new crow on the green grass ot the hill, fanning his black tail downwards and half opening his wings. Suddenly he made two great 15-foot leaps into the air, performed a few corkscrew flight gyrations, then landed back beside her, as if saying, 'Look what a magnificent athletic fellow I am'. And she, bent low, beak almost touching the earth, acted as if she were embarrassed and saying, 'Now then, laddie, no need to make a fool of yourself'!

Next morning both crows were in the birch tree, freshening up the old nest with twigs, beakfuls of dead leaves, and wool pulled from the carcass of a winter-dead ewe in the hills half a mile away. Hoping they would breed there, giving the place at least one bird family, I put out bacon rinds, bread and scraps for them each morning.

Just as in Canada, with the crow I had named Charlie, it became a one-way friendship. Crows everywhere are opportunists, quick to divine if a human is friendly or antagonistic, and before long the young male took full advantage. He too came early every day, waking me up like an alarm clock at dawn with his harsh '*Kaahs*' from the ash trees near the croft. I always stamped my food tins flat so they would take up less room in the wooden box outside before I buried them. And like the Canadian crow, if he didn't find enough scraps on the grass, he would haul out the meat tins, peck them loudly or throw them about with loud clinks, some of them landing in the burn. I named him Charlie II and he soon became a court jester, or rather croft jester, about the place.

In Scotland the hoodie is persecuted by farmers, shepherds, keepers and stalkers for alleged crimes such as pecking out the eyes of dying lambs, joining forces with ravens and black-backed gulls to prey on sickly sheep in winter, stealing grouse eggs and chicks, hens' eggs laid wild, or raucously warning deer of the presence of a lurking stalker. It is not surprising that the crow, which has no real song – just a big mouth – has learnt to use man for its own advantage. It is clever, arrogant, bossy and assumes the world belongs to itself – all such human qualities, so no wonder it's disliked! It trades man guile for guile, an optimistic knave which virtually thumbs its nose at attempts to eradicate it, and has even been known to spring the now-illegal gin traps for foxes by dropping stones on them so it can get safely to the bait. Several times in my first summer I saw hoodies sitting in rock shadows, patiently waiting for oystercatchers, ducks, meadow pipits and grouse to reveal their camouflaged nests, so saving themselves an energy-wasting search. They also acted as sentry for their mates, perching on high rocks while the others looked for the food. At first, of course, I knew none of this and felt strangely flattered that Charlie II found me trustworthy.

He was certainly a natty character in his jaunty grey waistcoat, and with his black wings and tail and his black cap, he looked as respectable as an undertaker.

Yet his Latin name gives a clue to his real character – *Corvus corone cornix* has a Mafia-like ring to it. It seemed odd that a bird which needs to be ultra cunning to survive should have such easily seen plumage, apparently totally lacking all camouflage qualities.

However, one old gamekeeper I met in the local village held no grudges against crows. 'Years ago, when we stalked stags, the hoodies would gather high above us in the sky, warning the deer, but that's rare today, though ravens will do it. I see far more crows on the seashore now than I used to. Perhaps they're changing their habits. Och no, 'tis herring gulls are the menace now, and if

crows are taking their food, 'tis maybe no bad thing. I don't
bother shooting hoodies now.'

In later days Charlie II became tame enough to drink milk
from a tobacco tin left outside, and it became his favourite tipple.
He would uptilt a few beakfuls, quibble his mandibles a few
times, wipe his beak on some bracken, fly to the big rock outside
my window until he felt like some more, then call peremptorily
for a refill. Crows are like that – give them an inch and they'll
take a yard. Like Uriah Heep, they work their way into your
confidence then take you for all you've got.

He also had a sadistic sense of play. I had once caught my
Canadian crow playing with a frog. Charlie II also ate frogs, but
one afternoon I found him fooling about with a toad. Toads have
poisonous secretions on their warty skins, so Charlie II had no
gastronomical interest in the creature. Instead he was playing
with it like a cat with a mouse. Every time the slow, stolid toad
tried to hop or walk away, Charlie seized it by a leg and dragged
it back, or else hopped round to block its escape. Pure horseplay.
Toads would be useful about the garden I intended to plant and
as I was certain it didn't share Charlie's sense of fun, I rescued it
by popping it into the burn. Charlie looked most put out that I
had spoiled his little game and flew off. He constantly reminded
me of the pet crow I had befriended as a boy after finding him
alone, apparently orphaned, in a big field.

Clark had possessed a devilish sense of humour too. Sometimes
when our pet tortoise Toby stretched his crusty old neck out for
a bite of lettuce, Clark would ding sharply on his shell with his
beak. Toby retracted his head like lightning and Clark stared
with fascination at this curious creature, walking round the
tortoise until it plucked up courage for another go at the lettuce
– then Clark dinged on his shell again. But a well-shied acorn
hitting him amidships, plus a spell of detention in his converted
rabbit hutch, finally cured Clark of this little prank. Then there
was our cat, a floppy good-natured, marmalade, neutered tom
called Ginger. For want of nothing better to do, Clark would

stage little fights with Ginger, the cat, usually when he was snoozing. Clark would spar about like a little black featherweight from the Bronx, bop the cat smartly on the nose with his beak or tweak his tail. Then as the furious cat swiped hopelessly beneath his rapidly rising form, Clark would fly to the safety of the conservatory roof and give vent to hoarse croaks that sounded like guffaws.

Crows have long featured in human folklore. In the Faeroes an ancient belief was that if an unmarried girl went out on Candlemas morning and threw a bone, stone or a lump of turf at a hoodie she would know her future marital state. Depending on whether the crow flew over the sea, landed on a nearby house or stayed where it was, she would know if her husband would come from over the sea, from that house or else she would remain a spinster. As the hoodie is usually a very wary bird and would fly off as soon as she even made to throw anything at it, she was at least certain to be reassured on the most important point. If the olden-day Scots said a man was a 'gone corbie', or looked like 'he would make the crow a pudding', it meant he was at death's door.

It was in early May that my friendship with Charlie II really ended. Like many jokers he began to overstep the mark. On several mornings I was woken up by loud thuds from the back of the cottage. At first I thought it was the usual sheep rubbing its head and horns on the slats of the wooden wall. Mystified, I crept into the rear room and found Charlie busily pecking out the new putty from the window I had installed. He seemed to like the linseed oil in it. That was going a bit too far. I opened the window fast and clapped loudly. After a couple of mornings of this treatment, he got the message and flew off to perch on a far-away rock, looking most pained at this uncharitable attitude of his new friend.

Charlie had provided some good laughs but the civilized side of island life had its amusing moments also. There was no bank for 45 miles, but my branch sent a mobile van to the nearest village once a fortnight, where it opened for business at 1 p.m.

and stayed for about 40 minutes. At first I tried to boat out to draw my shopping cash but once, delayed by bad weather, I didn't reach the spot until 1.30 p.m. No bank. I chased the van to a hamlet, and honked and flashed it down with the Land Rover lights,

'We waited for 25 minutes but no-one came,' the driver explained. 'But don't worry. We will bring the bank to *you* if you'll tell us where to meet you every second Thursday!'

Knowing the intransigent weather, I was never able to take advantage of this kind offer, and was finally able to cash my cheques with local traders.

This happened in late March, and on my way home I decided to visit Deer Island and spend a few hours in the hide I'd put up for the herons. While other herons seemed to be paired off in many of the other nests, the huge structure at my eye level seemed occupied by only one bird, a large male I presumed, for the males usually commandeer the nests first. Overhead a few lone herons soared and I was sure they were mainly unmated females for as each one came nearer to his nest, he raised himself to his full three-foot height as if to make himself conspicuous. He raised his head crest of black plumes and made two or three odd hooting or honking noises, while throwing his huge bill backwards in a kind of tossing motion. When any of the flying birds landed in his tree, he lunged at them with his beak and flapped his vast wings as if to frighten them away. With herons the sexes look alike, and as they are solitary birds most of the year, this apparent driving away of the first females is an instinctive action. It is up to her to prove her desire to mate, her 'love', so to speak, and persist. Yet not until I had spent three hours in the hide the following day did I see a female force herself upon him.

She flew first to two other lonely males, suffered herself to be driven away, then went back twice to the male I was watching. The second time his beak jabs seemed increasingly half-hearted; she evaded them coyly and he then appeared to accept her for he clapped the two parts of his great beak together noisily and gave hers a series of long gentle nibbles. Two days later he was courting her by bringing gifts of sticks, some about eighteen inches long, which he offered to her with an elegant display of his head plumes. She raised her plumes as if in thanks, took the sticks from him and arranged them about the nest, pushing and tugging them through twigs already there. Sometimes after this the two birds mated, she clinging to the precariously swaying branches while he mounted, both birds flapping their huge wings to maintain their balance. It seemed odd they should mate on such a flimsy perch when their feet were so well adapted for walking on the ground and standing firm on the most slippery of surfaces – rock, sand or plain old mud. I felt the female had a rather hard time of it for the male seized the back of her neck with his bill and held on to it while mating. At least it didn't last long. As with most birds a brief touch of the cloacae suffices to pass the seed and the moment is over.

By mid-April there were four large blue-green eggs in the nest, and it seemed that herons practise equality of the sexes. The male provided food for his wife, and for a month he seemed to take a fair share of incubating the eggs. Later each bird flew away after being relieved by its mate to find its own food on the seashore. Once, on a hot sunny afternoon, I saw both birds standing on the nest edges, shading and gently fanning the eggs with their great wings. While I sweated in the close dank atmosphere of the hide, I felt somewhat envious.

Twice, as I watched the heronry over the next few weeks, hooded crow couples flew close overhead. Once a hoodie flew near, as if trying to lure the herons off to attack, so that its mate following behind could dive in and purloin an egg. The herons I observed were not taken in by this ruse. Instead the female sat tight while her mate, perched on a branch nearby, pointed his huge beak at the flying crows, staring at them, his whole posture one of threat and aggression.

But clever *Al Corone* knows what he's doing, and when I returned in the sixth week and saw only three gawky, incredibly ugly fledglings in the nest, I could guess what had happened.

Sure enough, at the foot of the tree lay some smashed shell of the fourth egg. Probably a hoodie had swept down during a brief unguarded moment, had quickly flipped out the egg and had then had to quarrel with a crony or two for the privilege of gulping the contents. I felt that the herons had enough problems just then and decided to leave them alone for a while.

3

Dangerous Passage

After the Pacific coast years in Canada I had deluded myself into thinking that I knew enough about living with a small boat and the seas. Here I was on a long sea loch, barely three-quarters of a mile across at its mouth, and as the big spring tides rose and fell as much as seventeen feet, the tidal flow was fast. To avoid the barely submerged rocks, and gain help with rowing too, I learnt to keep to the centre when going out or coming in *with* the tides. When I went against the high tides I kept well into shore, for the back flow from the little bays and promontories reduced their speeds. There were three huge rounded rocks across the entrance of my own bay, easy to see at low tides, or to ride over on highs, but on medium tides their granite tops lurked just below the surface. Although I was advised to cut off the seaweed that grew on them, I decided against it. We clear too much of nature; seaweed is vital for oxygen-ating the oceans and as food and nurseries for many sea creatures, and I had learned in Canada that it can be useful as a cushion for a boat in rough waters. And as this weed always floated towards the surface it indicated the positions of the rocks anyway.

Nevertheless, it was on one of these boulders I lost my flimsy polystyrene boat. Coming back with a full load of paint, putty, nails, lumber and other bric-à-brac for the croft on a wet and windy May day, I was struggling to make headway with the oars when one of them snapped in half. A huge wave lifted me up – and down – *bouf.*

The keel hit the top of the eastward rock and, despite the seaweed, the boat split almost in half. Hurling what would sink

beachwards, I flopped out fully dressed, grabbing the bow rope
as I went. With the split hull now relieved of my weight and floating, I swam and towed it slowly ashore. Bedraggled and shivering
in icy water I rounded up what was floating but had to go back at
low tide for the gear that had sunk.

Although I repaired the oar by nailing the sides of a stout tin
round it as a splint, I was reduced to walking over the mountains
to collect my weekly groceries and mail which now had to be
brought to the big house with the Vanes' supplies. As I struggled
over the slippery rocky trail with full pack, constantly transfer-
ring a heavy 2½-gallon paraffin can from hand to hand, there
were moments when my friends' admonitions that I was too old
to live such a life seemed right. But above all I wanted to be totally
independent, so on my next trip south I called on a boating
company at Bowness in the Lake District and bought a pair of
magnificent ten-foot-long sculling oars for £3. Then I invested in
a 13-feet 9-inches Norwegian-style fibre-glass boat with a deep
keel, and a 7-hp outboard engine.

My bank account now almost empty, I paid a desperate call
on a leading popular magazine in the hope of selling them a
series on a major Hollywood star I had known well. At that
time Walter Clapham, author of several fine books including
Night Be My Witness, was serving as features editor. I knew
Wallie well from the time when he had been the film critic
while I was a cub reporter on the same Brighton paper. We
were discussing various ideas when suddenly he asked me why
I was now living so far away in the wilds of Scotland. Feeling
we were just wasting time, I told him briefly of my years in
Canada, of the grizzlies and the two old woodsmen I'd met,
and how I returned to Britain to see if one could actually live a
wilderness life here.

'I suppose I was looking for paradise,' I laughed before going
back to talk about other famous people I might write about.
Wallie listened, then he pounded his desk so hard with his fist
that his old pipe jumped out of the ashtray.

'Mike, you are a prize twit, if you don't mind my saying so! You're *sitting* on a good story.'

I groaned inwardly, sure he was going to ask me to deliver Sinatra by Sinatra, or something equally impossible like Brando by Brando, but I said meekly, 'What's that?'

'Your story, you fool. Your own search for paradise. Everyone wants that, but it seems you've actually *found* it.'

I stared at him. 'My story?' I repeated. 'I'm a journalist. I only write about other people's lives. No-one would be interested.'

Over lunch Wallie insisted I write him a synopsis, and I returned to the island, the Land Rover laden with the new boat, engine and enough planed boards to cover half the croft's front wall. As soon as I put the boat into the water I no longer begrudged its cost. The deep keel made it secure in the choppy seas and although, without the engine, it weighed 275 lbs, it was steady to row. I have always loved rowing and now I could not only save fuel on calmer days, I had a good replacement for city methods of keeping fit, like playing squash or exercising in gyms in artificial light, plodding round a park or flailing away on a rowing machine without water. From now on rowing would not only be good exercise but useful too.

It was low tide when I arrived back at my little bay so I grounded the boat on the soft sand, tied its bow rope to a big rock then began the usual heavily loaded walks up and down the steep slope. So busy was I for the next two hours that I completely forgot about the boat. At dusk I went down to haul it in on the high tide – only to find the rock to which the rope was fixed under ten feet of water! With a southerly gale springing up I had to stay awake until the early hours when the tide had ebbed again, to haul the boat in over constantly re-set planks to prevent it being damaged on the rocks when the high tide returned. Next day I invented a special cradle for it from ropes and three planks so that when the tides came in – I couldn't watch it *all* the time – the planks would protect the hull from the rocky beach. I thought of putting out a permanent mooring but on such a shallow beach, where the sea receded four hundred yards, and which was exposed to the prevailing south-westerly gales and heavy rains that would fill the boat up in a day, I realised that was out of the question.

I pounded away at the magazine synopsis. As memories of the Canadian cabin and the hard treks in the lands of the grizzly came flooding back, the story seemed to write itself. I suddenly realised that if I kept the main wildlife experiences out of the article I might have enough material left for a whole book. I shall always be grateful to Walter Clapham for thus nudging me into what became my first wildlife writing.

So excited was I by this new work that I was oblivious to all else until I found myself running out of stores. With the synopsis ready to post too, I had to go out. By the time I had fought my way to the little pier against strong easterlies, I realised the store would be shut for half day in twenty minutes. Dashing through a seemingly shallow pool, the water splashed up and doused the Land Rover engine, which misfired and broke down. I had to walk a mile and a half to put the synopsis into the nearest post box, then a further mile back to the pier before boating home. Once again the tide was low, and the pouring rain would fill the boat in a few hours, so I had to haul it up 350 yards over planks.

That day, unable to get stores, I began the first experiments with wild plant foods. All I knew then was that young stinging nettles were a nutritious substitute, if a little bitter, for spinach. But during that first summer I also learnt what other plants were edible by two methods, neither of which I recommend to anyone else. First, I watched what the sheep were eating – grass, wild sorrel, dock, dandelion and bramble leaves – and reckoned that if such were all right for sheep they probably wouldn't poison a human. Second, I ate only small amounts, then waited for signs of stomach trouble. If none, I increased my intake. In this way I found that in summer and autumn I could provide most of my vegetable needs from the land around me.

On fine days, when the sun became so hot through the south-facing window, I gave up writing and worked outside on the croft, repairing other windows, slatting on the new front wall-boards, then painting the exterior. One afternoon at the end of May I took a tin of red oxide and a brush and climbed up a heavy scaffold device I had invented to paint the iron sheets of the roof. Just as my head came over the top, I saw a huge brown-and-buff speckled bird hovering over the burn a mere thirty yards behind the croft. At first I thought it was an eagle, but eagles don't hover in still air, and its lighter plumage told me it was the big female buzzard that often flew overhead. Her great wings were winnowing the air like a huge kestrel, her body perfectly still as her large brown eyes glared at the ground below.

Suddenly she closed her wings, dropped with an audible thump, and stamped, making convulsive, griping clutches with the yellow talons of one foot, as if killing something. Then she jumped into the air and, seeing me for the first time, so intent had she been on what she was doing, she flapped her wings with a loud beating sound and flew up the hill with a large frog in her talons. I had never known a buzzard could hover in that way and it was clear why falconers use heavy leather gauntlets when handling such big birds. Not once had it attempted to use its beak; the killing had been done with the powerful talons.

Later, covered with dust and paint specks, I washed in buckets of water from the burn. Not *in* the burn itself, for I had no wish to pollute the pure waters. Instead I threw each bucketful of sudsy waste over the virulent bracken roots, so using it as a weed killer. By now the first deer flies, known in the Highlands as clegs, were vying with the little black midges to make outdoor periods a misery. If anything, the clegs were more irritating for they buzzed about slowly and noisily, landed right near the eye, in your mouth or on your back where you couldn't reach them. Often the first thing you noticed was a painful prick as they dug their hard mouth parts into your skin so they could suck some blood. Their slowness was deceptive for they often dodged your swatting hand, zoomed round in a tight two-foot circle, and landed back in the same spot for another try. While you might squash a few of them unpleasantly on yourself, a fair number got away with enough of your blood to help perpetuate their pestiferous kind. Mercifully, they seemed to come in batches, and there were periods when there were none at all.

I was sunbathing in one of these free periods when I was presented with a hilarious entertainment. As I looked at the stonework below the wall boards of the croft, I saw it was literally freckled with little dark spiders which had spent the winter in tiny crevices and were now out to catch the sun. They were jumping spiders which, while they can spin line to drop to the ground from danger, don't make webs to catch their prey. Instead they stalk and leap upon flies and other small insects like tiny tigers. I soon realised they were not just out for the sun or to hunt – they were also looking for mates.

The small black males outnumbered the larger lighter coloured females by some six to one, and if some human bachelors think romancing the female they most desire is fraught with troubles, they can be thankful they don't belong to the jumping spider fraternity. I watched one male attract the female of his choice. He advanced in short cautious bursts and quivered his left front palp at her. Then he paused and quivered his right palp. He paused again before quivering both palps at her intensely. Five times he did this, and it seemed that each time she received the vibrations from both palps she made rushes at him. Each time he dodged back slightly, showing submission or at least respect. She then ran off some four inches, stopped as if she had changed her mind and went back to what seemed his irresistible appeal. Again he went through his routine, moving nearer, but then another male with only seven legs came along and had a brief skirmish with the first. The fracas lasted less than half a second, a quick flurry, then both males fell off the wall, a mere two inches. The first spider dashed into the grass and keeled over dead while the one with seven legs staggered about in two-inch circles looking very shaky on its pins. In that brief time it seemed poisonous bites had been delivered by each spider. Others were quick to take advantage, however, and one small male now made a direct approach to the female – to be literally heeled away by an irritated flick of one of her rear legs, like a fencer not even bothering to look where she was thrusting. It seemed she had no time at all for a potential beau that hadn't bothered to learn a true palp technique.

Two more males received the same treatment then, as if in disgust, the female ran down the wall and into the grass. Within five seconds a new female had occupied her former space. One male began to court her with his palps while another dropped down the wall from the boarding crevices, attaching its line to the stonework as it went, and landed one-and-a-half inches from the female. Instantly she attacked it and retreated again, two inches to the left. This bitten male also fell to the ground and began to stagger in circles, his movements becoming slower and slower. Now the seven-legged warrior, having recovered from the bite by its male opponent, rapidly scaled the wall towards the new female. He was just in the middle of a new palp routine when the male the female had bitten, seemingly recovered, re-entered the arena. This time he too got the worst of it in a brief fight with Seven Legs and once more fell to the ground. Seven Legs now appeared to feel he had more than proved his valour and had no more time to waste. Minus a leg or not, he certainly knew his palp technique and started making intensely violent movements at the hypnotized female – left down, right down, left up, both down and quivering. Again and again he repeated this odd sema-phore until the female, seemingly convinced he was a worthwhile mate, walked slowly towards him. She did not present her rear. She didn't have to, for with a great leap Seven Legs was upon her and, clasping her tightly, allowed her to hurry off with him into the grass below the wall. No battlefield Hector deserved more the spoils of war.

What impressed me most was how the spiders could leap four or five times their own length on a *vertical* wall. The acrobatic techniques for such a feat are mind-boggling, for the spider has to leap out slightly and move forward, yet pull sufficiently hard with its upper legs so as not to lose height, stick in its legs while in flight, then extend them at the end of the leap at just the right moment to grab the wall again. Yet they could do this several times in a second and still attach the line before each jump. Astonishing feats indeed.

By now the seals which hauled themselves out at low tides on to the rocky islets offshore, to rest in the sun after hunting fish, were well used to my comings and goings, and one afternoon as I rowed back from the pier they started to follow the boat. They lifted their puppy-profiled faces from the surface, looking at me with lazy eye blinks, their nostrils opening and closing for a hissed breath, then somersaulted down again with violent rear flipper splashes close to the moving boat. I found when I sang a song to them – such as *If Ever I Would Leave You* – they came even closer, and by the time I reached my bay there would be more than a dozen wet gleaming heads bobbing up and down a few yards away.

Once, on returning, a pair of beautiful black and white oystercatchers flew up from the beach and began diving close overhead, issuing shrill '*pleep pleep*' alarm calls from their bright orange beaks. As soon as I landed they became silent, just flicking by with shallow wingbeats, reminding me of woodcocks in 'roding' flight over their woodland territories. I felt sure they had a reason for this strange behaviour and walked over to where I had seen one of them rise. I almost stepped on their nest. Instead of making it amid the small carpets of herbage above the beach they had scraped out a shallow depression in the shingle lower down, almost as if to camouflage their three greenish brown-blotched eggs, for at a few yards they were indistinguishable from the multi-coloured pebbles. Sheep often grazed along the shore line, and no doubt in this way the oystercatchers avoided their nests being trampled. They were wily birds, for while their noisy 'mobbing' had been to discourage me from landing too close, when I was actually very near the nest, instead of showing its whereabouts by increased alarm, they had fallen silent. From then on I pulled my boat up on the far side of the beach.

I was working at my desk one afternoon when I saw a flock of gulls spiralling over the shore hills to the west. Some were diving at a pair of buzzards and shearing away again at the last moment. They seemed to concentrate more on the bigger female, and when I went out to look I found out why. The smaller male was more adroit in the air and seemed finally to become angry for it turned on one gull and chased it a few yards, forcing it to swerve wildly to escape. Well, gulls do not spiral for nothing, so I hiked over to see if they had found a deer or sheep or lamb carcass, but a thorough search revealed nothing.

I came back along the beach and once more found myself in the secluded little glen right on the shore. As I looked at the square formed by the six different kinds of trees, all now in full leaf, I thought what a wonderful place for a small cabin, a beach studio from which I could watch the nesting eiders, gulls and oystercatchers on Eilean an t-Sabhail, a small islet just off shore.

I could probably build the whole thing behind a sheltering rock slab so it would be invisible except from directly in front, where few boats passed.

With my thoughts back in Canada now that I was writing the outline for the book about life there, I felt strangely nostalgic. There is something magically satisfying about building with logs, scribing the knotted, rounded, uneven timbers so each fits precisely into the other, about making a home from natural objects. It also suddenly seemed important that I prove to myself I still had some of the skills taught to me by old Ed Louette.

That day was my birthday, and on the way to buy a celebratory bottle, I met Digby Vane who invited me back for drinks at the big house. As we talked I slipped in my request to build the little cabin. They had now decided I could stay as long as the island remained theirs. I could indeed build a cabin on the beach, and what was more, if I ever had to leave they would repay my costs, but my labour, being self therapy anyway, would be free. I determined to cut no trees down, to use only windfall logs which otherwise would rot. While there were a few in Shona forest, they would be hard for one man to drag to the beach. I had noticed there were many on Deer Island which I could cut and tip off the low cliffs right into the sea for towing. I contacted the island's owner who said I was welcome to take a few fallers for a cabin.

Working mainly on windy midge-free days, I toiled at the cabin site, grubbing down to bedrock for the corner posts and digging a peaty-water well on slightly higher ground so I could siphon water down. When I boated to the pier to fetch cement, timber and a pair of heavy oak doors – bought for fifty pence each from a demolition site on my last London trip – the easterly winds freshened to slight gales. I was loath to wait and come back another day so, taking a chance, I set off.

As any boatman knows, every engine and boat combination is different and I was now running fully laden before heavy seas. The speed balance had constantly to be altered, for if the boat went too slowly it wallowed between the troughs with the waves threatening to break over the stern. If I went too fast it slid down the far sides of the troughs with a great rush, nearly ploughing the bow under the swell of the wave in front. So I had to keep a little faster than the waves when going up them, then throttle back when shooting down again. It was a rough trip, filled with sun and spray, but I reached the shelter of the cabin bay safely.

Leaving the boat out at anchor, the tide well down, I managed to haul the lumber and heavy doors up to the site, stumbling thigh deep in the slippery rocky pools, and after a further hour of work felt very tired.

Some fifty yards east of the site was a large, flat, sheltered boulder covered with thick moss, now dry from sun and wind, and I went to lie down there for a while. How wonderful that moss seemed, a thick, green carpet which softened the sharp contours, laying a curving, caressing touch to the tired body as if it were provided by nature for a man to rest from his labours, a lush, miniature forest, cool to the touch, in which few insects dwelt. I lay there looking into the cobalt vault of heaven, softly combed by the light-green leaves of a hazel bush that shimmered above me in the breeze, and fell asleep.

4

Rescuing a Dagger Bill Heron

Suddenly I woke up to strange bleating cries coming on the wind. Keeping as low as possible, I crawled to a small heathery ridge some twenty yards away and peered through. It was an extraordinary sight. A large browny-red hill fox had cornered a young lamb and appeared to be playing with it. As the lamb bleated helplessly, the fox darted underneath it, forcing it upwards and off its legs, jumped over its head, pushed it down with two black-tipped paws – enjoying itself like a cat with a large, white mouse. Before it could do anything more I leaped up to scare it away and it ran, long brush streaming like a thick banner, and disappeared between the rocks. The lamb continued bleating and scurried away. It seemed strange its mother was nowhere in sight but I found a large ewe some forty yards away to the windward, placidly grazing behind a ruined wall that obviously had not allowed the lamb's cries to reach her.

Now the lamb ran to her with piteous bleats for a reassuring suck but the ewe became alarmed and at first tried to butt it away. Then she seemed unsure of what to do, whether to accept it or keep driving it off. Evidently the fox scent tainting the lamb was the trouble for she kept huffing and snorting, torn between maternal desire and fear. When the little lamb persisted and she got the right scent too, she finally accepted it back, and soon trotted off like an animated mop with her youngster wobbly on its legs beside her.

To see a Highland fox in broad daylight was rare enough but to see it actually *playing* with a lamb, perhaps before the kill, must have been a thousand-to-one chance. Certainly I have never seen the like since.

One day, wondering if a huge fallen tree I had glimpsed in the Shona forest might be a red cedar from which I could split 'shakes' (wooden tiles) for the cabin roof, I walked the hilly track only to find it was a silver fir, useless for splitting. I searched in vain for a fallen cedar but there were compensations. A brown blur emerged from a mossy crevice as a robin left her nest, revealing four white and pale red-speckled eggs. Further on, a short-tailed vole chewing young grass suddenly became aware of my looming shape, zig-zagged about a rock then spying its hole, dived down it. Green-veined white and speckled wood butterflies were flitting in the clearings and I saw a first green hairstreak, sunning itself on a leaf, a jade jewel of the forest.

Next day, feeling like an old-time hand logger, I boated to Deer Island with a scaffold jack, a car jack, felling and hand axes, and my four-foot tree saw. To procure big logs from a forest on his own, the hand logger finds or cuts one on a slope above the sea or a river. He cleans off all the branches, then cranks the jacks under the high end to tip it downwards. With care and luck the log heads into the water but it also often gets hooked up so the process has to be repeated. I climbed the steep cliff on the west side of the island and got out my first two fifteen-footers. One went down superbly, hitting the calm water with a great splash. The thicker log hit a rock, somersaulted and landed amidships across another rock, splitting in two.

I 'timber cruised' the edges of the island for more logs, knowing I couldn't get any out from the centre on my own. The heronry was now in full spate, the great birds wafting above the tree tops like huge snowflakes. I didn't want to disturb them. After checking that the three fledgelings in the nest I had observed earlier were now sprouting feathers, I went to the north-east edge of the island. There, bridging a wide, marshy ravine, was a fine, dry larch tree just over a foot thick. Using the quiet hand saw (I couldn't afford a chain saw) I cut first one end, then half through the other, to drop a 30-foot log down far enough so that I could cut through its centre without risking injury or death. As I dragged each 15-footer along the marsh's edge, where the mud was not too deep and down to the sea, I was surprised by how heavy the logs were. Then I realised why. Western Canadian trees can grow to a foot thick at fifteen to twenty years old but these larch did not attain that thickness for about sixty years, hence the wood was denser and heavier at the same size. Cutting and dragging once more, I heaved them into the sea, rounded up the floating log and towed them all to my own beach four miles to the west.

As we laboured into the open loch past Riska Island, I saw a dull-brown eider duck shepherding a flock of eighteen ducklings. The little ones scattered, racing over the water with flailing feet, their flipper-like featherless wings beating unavailingly, like little seaplanes trying to take off. The mother fluttered before the boat, trailing her wings to decoy me away from the babes. Then she dived to emerge among them many yards away. They could not all have been her own brood, and this and later observations led me to conclude that an eider mother will sometimes take over another's brood for a time, perhaps allowing the second duck to feed at some choice place. Old or unmated ducks will also help shepherd ducklings. It is at this time the youngsters are most vulnerable, for they are preyed upon by the big, black-backed gulls, can be seized from beneath by seals or otters, and on land by foxes, wildcats or even rats. The great skuas also eat them, harrying the brood which will dive and dive until exhausted. Skuas then take them easily. The skuas lived further north, and there were none on Loch Moidart.

That night I felt too tired to haul the logs up to the cabin site so I left them tied by a long rope on my own beach, going down in darkness later to haul them in on the high tide. In blazing sun and tidal calm I towed them next day to the cabin beach. Now wet, they were heavier than ever, and I soon cut my hands trying to drag them with short heaves over the rocks. I was assailed by midges but I foiled them by tying butter muslin round my face. I could hardly see through it, sweat poured down but anything was better than those continual bites.

Even the harsh grey granite rocks themselves seemed to possess demoniac characters of their own. Timeless, enduring, mocking the puniness of a labouring man, still and dead on cold days, treacherously slippery in the rain, they now shimmered unendurable heat waves into the air, sapping my strength with their reflected heat, baking and cracking as if biting back at the sun. Anchored motionless for ever as they seemed to be, they were yet able to destroy, as if harbouring malicious spirits that forced my feet the wrong way as I toiled uphill with the logs, twisting my feet in their crevices as if they had planned it all for my hardest tasks. What a difference it makes to have a few inches of soil and soft grass beneath our feet.

I de-barked the logs with the hand axe, cut the wooden corner blocks, and set the foundations with cement, sand and granite chips. I was soon in trouble. The cement was too crumbly from too much sea salt. I had to boat to Shoe Bay for fresh sand from well above high tide level, where all the salt had been leached out of it by years of rains, and then re-make the foundations.

After cutting and hauling three more logs from Deer Island, I was busy heaving them over the rocks with the rope sling when I saw the boat floating away. Anxious to get them on to land while the tide was still high I had forgotten to tie it up and had underestimated the speed and pull of the ebbing tide. Now the east wind was blowing the boat westwards. I tried to reach it with a pole, but it was now too far out.

I raced over the beach to the last land point, but again it was tantalizingly just out of reach. As I watched it heading beam-on for the great Atlantic I realised I was about to lose both boat and engine for good. Off came my clothes. I dived in from the rocks, got my underpants tangled round my feet, kicked them off, lost them, but managed to reach the boat in forty yards. As I tried to clamber into it stark naked I heard sudden shouts. A boatful of young Sea Cadets who were camping on the north shore of the island, had chosen that moment to make their first and only boat trip across the mouth of the loch.

'Do you need any help?' they yelled.

'Nope,' I growled back through chattering teeth, my nakedness modestly curled up under the keel.

I returned to the croft to discover that in my hurry to haul that day's logs and the boat off the mudbanks at Deer Island before low tide I had left behind my hand saw and thickest boat sweater.

When the rain eased off next morning I saw two chaffinches searching for food on the big mossy rock below the window. I quickly made a small bird table from a piece of plywood, with upraised edges so that food would not blow off, put on a crust of bread and returned indoors. Within seconds the colourful pink-chested cock bird landed, took a beakful, swallowed it, then seemed so happy that, with both feet firmly on the bread, he burst into his little chippy song! As he chirped away, full of cocky health, the three white stripes on each wing blazed like chevrons, and I coined a name for him: Little Fat Sergeant. I had never before known a chaffinch to sing from anything but a tree branch.

After sudden overnight gales I woke to a calm sea, the air hazy with occasional sunny patches. I boated to Deer Island with the usual nails, rope and axes, to recover my saw and sweater, and to fetch more logs. Before starting work I went to check the hide on the knoll and the nest of young herons. As I tried to walk unobserved beneath the tall conifers and the vast nests, some of the birds were with their half-grown fledgelings. But with their acute senses – they can hear a twig snap on the forest floor and have slightly forward-placed eyes incorporating almost the stereoscopic sight of falcons – they knew I was there. They froze like grey bitterns, while dozens of yellow eyes, bright as tiny catherine wheels, glared down at me from long stalky necks. I crept into the damp sacking of the hide and focused my field glasses. There was only one cowed youngster still in the nest.

The overnight gales had uprooted a pine which, in falling against the nest tree, had dislodged the other two nestlings. Then something grey below caught my eye. On the ground under the tree a young heron lay dead. Hanging from the twigs of a thick bush by a large foot was the third nestling. At first I thought he too was dead but then I saw his neck slowly bunch up like a snake's, in agony. Then his half-grown wings flapped feebly as he dangled. He was clearly not far from death.

I climbed up to release him but as I carried him down, weak and cold though he was, he raised his spiky head crest in terror, trying feebly to stab at my hands with his beak. Placing him in some sacking, I hurried to the boat where I had a small fish I had caught by trolling on the way over. Although I knew herons feed their young on pre-digested food by regurgitation, I felt a little of its flesh would do him no harm, and after mashing it up with my fingers, I pushed a few morsels down his throat. Then I had a good look at him.

He was a ridiculous, misshapen, ragamuffin character, his huge feet, head and long beak completely out of proportion with the rest of him. Laughing at myself for feeling the need to give him a name, for I dislike bestowing human traits on wild creatures, I called him Harry. (Long since I have come to realise that only after a creature is given the dignity of a name, spoken in loving terms, can it respond to a human being without instinctive fear or fight. Maybe it is also one's own attitude that changes with naming, and that is what brings the response, but the effect is undeniable.) I checked him over for injuries. All seemed to be well except the outside toe of the left foot, which was completely broken and dangling loose. Luckily the important third toe of each foot was undamaged. These toes are vital to herons. The claws are like serrated combs, and are used for cleaning slime from eels and other fish from the feathers and to clear away body pests. They also help to hold a fish down. In preening, the claws help to spread a waxy, whitish-blue powder from six special tracts on the body over the feathers to give them a waterproof sheen.

Harry's toes were not webbed like a duck's but adapted for both perching in trees and working in marsh or water. They had spongy, spreading under-surfaces, ideal for walking on soft mud in shallows or over slippery rocks. His long beak had jagged edges for holding fish and his gullet seemed almost as elastic as a pelican's.

At first I was tempted to take Harry home but then I realised he would still be needing his parents' pre-digested food. Besides, it is better to leave fledged nestlings for usually their parents are still feeding them on the ground. When I saw a heron flying low near the trees, I placed Harry in a prominent spot, made a show of leaving, then crept back to the hide. For two hours I watched poor little Harry crouching there but the mother heron only landed by him once. She was clearly perplexed for she could not possibly get him back to the nest. When he tried his nestling trick of grasping her bill to induce her to feed him, she leaped gawkily into the air and went back to her remaining fledgeling in the nest. Clearly if I left him there his parents would abandon him and he would die. There was only one thing to do – try and put him back in the nest myself.

With Harry tied in some sacking round my waist, I drove six-inch nails into the tree trunk where there were no branches and, using a rope also, managed to work my way upwards. I was astonished to find myself struck by a severe attack of vertigo, something I had suffered from slightly since a fall from the cliff top in Canada. At times I had to cling to the trunk, shivering with dizziness, and fight a strange desire to just let go and fall. But talking loudly to myself, I went on. Near the top climbing was easier because of the extra branches of the lodged tree. Even so, it was hard to reach round the four-foot width of the nest and only after being covered in dust, decayed vegetation, heron pellets and smelly undigested fish bones, could I manage to push Harry into it. Climbing down again seemed even more hellish as I kept removing the hammer from my belt to pull out the nails as I descended.

Shivering with fatigue and fear, I was in no mood for more logging. After finding my lost saw and damp sweater near the muddy edge of the lagoon where I'd launched the first logs, I boated home.

A few days later I returned to camp out and do a two-day shift getting all the logs I needed for the cabin. As the boat putted past the island there was a commotion among the terns nesting on the islets off the north-eastern edge. As I drew nearer I was surprised to see a large tawny-grey animal with a thick, dark, striped tail bound off the rocks into the water. As it swam rapidly towards the shore its pointed ears flattening down as if to ensure water would not get into them, I realised it was a wildcat. It swam strongly, head and shoulders well clear of the water. It covered the 15 yards in less than half a minute, sprang out and, without bothering to shake itself, bounded over the rocky bank and vanished into the woods. I was sure it had been after the young terns. Leaving the birds to settle, I beached the boat and walked through the trees to check little Harry Heron. He was still there. After a short while the mother returned and Harry started to compete lustily with the other nestling for the food from her bill.

That night, after cutting more logs in strong winds, I camped well away from the heronry. In the morning the first thing I saw was a tiny brown wren emerging from a hole in a hollow larch snag. I thought little of it; probably it had wanted a warm berth out of the night winds, but then out popped another, then another. In the end thirteen wrens came out of that hole, rather like morning commuters emerging from a tube station in the city. Wrens often share holes in trees to keep warm in harsh winters, though the bird is usually warlike towards its own species at other times of the year. It seemed odd for at least two families to have shared that hole during the summer night.

As I slid out of my sleeping bag beneath a staked-out plastic sheet, the wrens appeared to gang up on me, all lining up on the twigs of a bush and making their chirring alarm calls at this odd newcomer to their world. After a sandwich breakfast I crept around the island but failed to get another glimpse of the wildcat.

I was de-barking a log in the afternoon when I heard another commotion from the tern rocks. I hastened through the trees. Now four herring-gulls were showing interest in the nests, and the smaller terns were swooping and diving at the wheeling gulls, crying out loudly. One tern was still on the rocks, its scarlet beak agape as it defended two tiny grey youngsters from one of the gulls. I launched the boat, and as I drew near one of the fledge-lings tottered, half-blown by the wind, to a cleft on the edge of a high rock. I feared it would be dashed to death, or drown if it fell into the sea, for as yet it had no flight feathers. I rowed in, rescued the little mite and restored it to its nest in some grass tufts on top of the rock. It showed no fear at all. The loud shrill screams of the adult birds, hovering and swooping above, seemed to die down a little, as if knowing I meant them no harm. When I rowed back to pick up my logs, they settled down on the islet again, folding their long, swallow-like wings and running to check their young ones.

I boated home with the logs behind me. Thousands of tiny elvers could be seen in the clear water below, all shimmering like sequins in the golden sunlight as they neared the end of their incredible three-year-long 3,000-mile migration from the Sargasso Sea, where they had been born. Now they were heading for the fresh-water burns, rivers, lochs and lochans up to 1,000 feet high, where they would spend up to fifteen years before turning into mature silver eels. Then they would set out once more, their stomachs atrophied for they would never feed again, to return to their birthplace, where they would breed and finally die.

Once again, as I reached open water, seal heads began popping up around the boat as they followed me home. I'm sure such animals, so well adapted for catching their fish prey, often feel bored. And when they become used to a harmless figure in a boat, they follow him just for fun.

As I lay in the sun by the cottage to rest awhile in the last rays before it sank below the high hills, two grey wagtails with pristine yellow breasts appeared from nowhere and began running over the ground, snapping up tiny insects. And a few yards to my right I heard a humble-bee drumming in a foxglove flower.

How good it would have been to share such moments with a beloved woman, and it was often in such happy periods that a feeling of intense loneliness would come upon me. I wondered if beauty of landscape could really be sufficient in life without human love. The thought made me sad, for love had looked in my door many times during the years I had been caught up in the international set in the cities of Europe and North America, and I had known some of the world's most beautiful women. Now, for the first time, it hit me that I was going to have to face the future totally alone.

Suddenly the awful loneliness of the first winter in Canada came flooding back. Alone in the wilds, a man's conscience and self-awareness increase, hang over him like a hawk. The psychological effect of one's opinion of oneself can be overwhelming in its ability to destroy – and also to preserve. Self-respect is the only foundation upon which one can survive. The new realization that I was almost certainly going to face the rest of my life totally alone seemed frightening. Surely, there is someone for everyone?

The point is that one is not lonely for just *anyone*, but for that special person whose mind inspires respect, whose interest comes from similar experiences, whose physical being attracts, whose spirit is parallel to one's own. I knew many men caught in a mutual trap with women they no longer loved, with no chance or energy to start again. Loneliness for one seems better than a spiritual jail for two. Alone there is hope.

On my next trip south, I worked up a few small writing jobs, bought materials for the cabin – glass, window frames, plywood for the floor – and on my return added to the full load with rough-sawn timber for the walls from a saw-mill near Fort William. Hard labour has always been for me a therapeutic way of alleviating loneliness. I ferried most of the materials in one trip by towing the timber.

As ever, wildlife-watching provided compensations for the lack of human company. One afternoon I was jointing the floor joists into the foundation logs when I saw a tern dive straight into the sea after a fish. It did not come up again. I waited, wondering what had happened. Suddenly there was a violent splashing – and a large sea trout leaped out of the water and grounded itself on a smooth rock, just as a seal's head emerged and turned with a swirl. Thinking I could get an easy meal, I ran over the rocks, slippery from spray. Before I got near the flapping fish, it flopped back into the water and disappeared. Nor did I see the seal again. I wondered if it had caught the diving tern or if the bird had become caught up in some seaweed. When I boated over I could not see it anywhere in the clear water.

Occasionally I went to Deer Island to see how Harry Heron was getting along. By the time he was about seven weeks old he and his mate seemed as large as their parents. Harry now climbed to the outside branches well away from the nest and, clutching desperately with his feet, using his long beak as a stabilizer, exercised his wings with violent movements. At times he flapped so hard that the branch shot up and down as if about to snap. He seemed to have an innate building instinct for occasionally he flapped and ran over the foliage, as if afraid to trust his wings alone, snapped off a small twig and brought it back, as if trying to help his parents repair their nest.

When I went back in late June I found the nest empty. When I took down the hide and walked out of the trees, there was a great flurry as a whole mass of herons suddenly flew up from the rocks of their spring gathering ground. The sky was literally thick with herons, and I counted thirty-one as they dispersed to the meadows and loch shores. It was an astonishing spectacle, exhilarating too, for it was obvious to me at that moment that if such a huge sedge of the big birds ever decided to attack a lone intruder in a boat, he would have a hard time against all those pickaxe bills.

Some of the birds were flying awkwardly, certainly this year's young, and as I scanned the sky with my field glass I saw one of them had a broken toe that stood out clearly. I was sure it was Harry.

I doubted that he would survive for long. Only one out of every five herons who leave the nest ever reach breeding age. Harry had a broken toe and had nearly died after falling from the nest. And while some adults stay with their young for a time to show them how to catch prey, many young herons have to look after themselves. Harsh winters, the chance of choking to death on furry prey that hasn't been dipped properly in water, colliding with overhead wires, having their huge wings broken in storms, falling victim to man's pollution and pesticides, or being shot by fishery bailiffs who regard them as poachers – all take their toll of herons. In any case, a heronry can absorb only about 20 per cent of the youngsters of breeding age – the rest have to find homes elsewhere. It seemed unlikely that Harry would become the one heron in five who survived.

But one late summer day, when I was boating past a small rocky islet near my bay, I saw a heron standing on one leg on the line of tide wrack. Remarkably it remained still until I was 30 yards away, then it rose with the usual gawky leap. There, outlined against the mackerel sky as it folded its legs neatly below its tail, I saw the broken toe again.

He seemed to have adopted my own beach for hunting his fishy prey and I began to see him more often. Sometimes, when I was leaving in the boat, he would fly over quite low with hoarse and strident '*kraink kraink*' calls, then glide to shore with soft chuckling noises, stick out his long gangly legs and land as lightly as a toy balsa-wood glider.

Like all herons he seemed to have two main hunting methods. The first was 'stand and wait', which he would do with the utmost patience for as long as two hours until some unwary little flatfish, eel, shrimp, crab, frog or vole came within his reach. The second was 'sneak along slowly', like a cat stalking (but knock-kneed) and with his head jerking back and forth like a ludicrous metronome. Whichever technique he used, the end was usually the same. A lightning dart or two with his beak, then he uptilted the prey and swallowed. Once I saw him beat a small fish against a rock to kill it first.

He peaceably allowed other smaller birds to feed alongside him and he often worked the shallows along with curlews, ducks, gulls and oystercatchers. While I saw occasional large gulls torment him by swooping low after he had caught something, as if trying to make him drop it again, I never saw him harry them. It seemed his attitude was, 'There's room enough for us all.'

5

No Compromise

To many, a life of isolation on a Scottish sea island would be a recipe for boredom, or worse. 'What on earth do you *do* all the time?' I was often asked on supply trips. In fact, there weren't enough hours in the day or night for all my tasks and I never got around to piping water into the house from the burn. I remained a 'three bucket' man through the years on Shona – one for washing water, another of sea water for cooking, and a bigger one for slops which were emptied over the bracken. A plastic two-gallon container held my drinking water. Two driftwood shelves carried my books, small battery record player, letters and wildlife files. My ornaments were abstract 'art works' fashioned by the sea into weird shapes from the roots of trees. The old hair mattress and bedframe rested on sea-scoured fish boxes found on the beaches. My bedside 'cupboard' was also a fish box.

Until my garden was producing, I fed as much as possible from the land around me. While using nettles for spinach, dock leaves for cabbage, dandelion and sorrel leaves for salads, and blackberries, hazel nuts and rose hips as fruits, I also conserved fuel by placing one pan on top of another when cooking. Thus the steam from boiling cabbage, carrots or nettles would also cook the potatoes. When making a stew I also heated the washing-up water by putting it on top of the pan instead of the lid. Saving water in which vegetables had cooked, I could add sliced onions, some lentils and stinging nettles, dock or sorrel for a first-class vegetable soup. Minced beef, fried lightly in onions, or lamb's heart stuffed with sage and onions, could be simmered with split peas and lentils, carrots and potatoes, or with any of the wild foods.

In place of a refrigerator I used a plastic box kept cool in a draught by the evaporation from a wet cloth. A waterproof galvanized box was better still when placed in the cold waters of the burn, though it needed tethering with light rope. Once, after a spate from heavy rains, I found my meat fridge a few hundred yards down the burn, where it had burst open on a flat rock. Charlie Crow and his mate rose into the air with harsh '*Kaahs*' – and the last piece of meat!

For a bread- and cake-making oven I used a large biscuit tin inside an old cut-out petrol can which was placed on a wire grill above the open fire. In the can and around the tin I put a layer of flat stones for insulation from the direct heat of the flames, then set the dough on greaseproof paper in the biscuit tin. I covered the lidded tin and can with a curved, iron sheet, which threw the hot air back around them.

I also fed from the sea. It was no trouble to take my rod and row out past the islets into the open Atlantic and sit fishing in the sun. It seemed ironic that I could live healthily and happily then on £5 a week yet spend much of the time doing what I used to cram into three weeks' holiday a year. I fished, rowed, swam, sunbathed, walked the glens and mountains, fitting work into the pattern of the weather – all part of a natural life rather than a hurried once-a-year escape into the outdoors when it took time to wind down, when city life had so blunted the senses one just could not see, feel or hear many of the inspiring subtleties of the wilderness and all its wild life.

I soon found fishing in Scottish waters totally different from the fish-rich seas of Canada. No Atlantic salmon would take a lure trolled behind the boat as they had in the Pacific. These fish were filled with stored energy after years of sea feeding and were now intent on going up the rivers and fresh-water lochs for spawning. Here the seas seemed emptier of fish and it took time to learn new techniques. But there were mussels by the thousands. I farmed them, selecting only a few big ones here and there so that the colonies would propagate naturally. The little black winkles were common and there was usually a fair spattering of chubby cockles left on the sands after storms had stirred up the sea bed at low tides. Then the big spring highs washed them up the beach, to be exposed on the next lows.

I used mussels and cockles, half cooked to make them firmer, as fishing bait, holding them on the sandy bottom of the sea with a sliding weight on the line. If it was stormy I would leave the boat and cast out from the shore rocks. But I often had a lengthy wait for the bottom-feeding plaice and dabs until I learned to ground bait the area with some crushed mussels. I discovered that, unlike in Canada, the bigger local crabs clung tenaciously to the bait with their claws, reluctant to let the easy food go! I could haul them all the way up to the surface and into the boat before they dropped off, making me snatch my naked toes well out of reach of their tough pincers. I was glad now I had not cut the broad leaves of the tangleweed from the three main rocks for on the lowest tides – when four hundred yards of sand would be exposed – I found the biggest crabs used the weed to keep wet and as shelter. Although crabs can breathe under water or on land with equal ease, they do need to keep their gill chambers damp. Again, I took only a few of the largest, leaving any less than four inches across the shell. Herons, gulls, hooded crows and oystercatchers also used these weed shelters as a food source when the sea was right out.

I became almost as dependent on the tides for food as the wildlife about me, learning that the highest and lowest spring tides occur when the sun and moon are in line on the same side of the earth, so exerting a combined gravitational pull on the seas. And that the Neap-tides occur when the sun and moon are on opposite sides of the earth, their separate pulls tending to cancel each other out. Springs and neaps alternate roughly every two weeks throughout the year, low tides follow highs roughly every 6¼ hours and each takes place some fifty minutes later each day. Island life makes it essential for one to adjust to these natural rhythms.

Slowly I learned new tricks for coping with the boat in the winds and tides on the long, shallow beach. The Vanes gave me two heavy-iron disc anchors to set from the boat's stern. Instead of re-setting planks and laboriously heaving the boat over them to beyond high-tide reach, I could now leave the boat on the sand on her rope-and-plank cradle, go down at high tide, give a few heaves and in she would come.

Leaving the boat at the little pier on the mainland gave me problems at first because the flow from the river Shiel on an ebbing tide tended to push the boat into the stonework. I always dropped the rear anchor a few yards out, leaped out of the prow on to the pier with the bow rope, then walked up and round to the side on the land, thus swinging the boat outwards again before tying up so she would not foul the pier. One day the island's head worker, Iain MacLellan, watched me with a slight smile.

'Now then, Mr Tomkies. There's a better way than that. Do you not know how to flick the *cruaidh* off the prow?' I said I didn't, and what was a *cruaidh* anyway?

Iain explained it was the Gaelic name the old-time crofters used for an anchor, usually a heavy stone tied to a rope. He agreed that the new disc anchors were an improvement, for with their centre holes they were easy to handle, and their rounded edges meant less wear on the rope from the sea bed. He now showed me how to 'flick' them off the boat. Balancing them so they teetered on the edge of the prow, and holding the bow rope in loose coils, he gave the boat a long, steady push out to sea. When it had gone some thirty yards he gave a violent jerk on the rope, the anchor discs tipped over and dropped to the sea bed, holding the boat far further out than by my method.

'Well, I'll be . . . That's a good trick!' I said, wondering how I'd got through the Canadian years without thinking of it myself. I looked at Iain MacLellan with new respect.

He was not a tall man but nor was he short. He had quick, bright-blue eyes, a high sensitive forehead, an urgent almost huffy way of speaking and was as much at home with the Gaelic as with the English tongue. Although at first he appeared slimly built, his shoulders were square and straight, and as he worked the big boat, loading stores and heavy equipment for the island with quick, sure movements that belied his fifty years, it was clear he was as tough as whipcord and as strong as an ox. Iain had been born on Eilean Shona, and apart from a wartime spell in the Scots Guards, when he'd been a highly regarded piper, he had lived there all his life. The fact that I had once been a regular in the Coldstream Guards gave us a little in common, but with the Highlander's typical reserve, it was more than two years before he called me anything but 'Mr Tomkies'.

One August evening Iain suddenly appeared at my door. He had lost Queenie, a new border collie bitch he had bought for £6 from a mainland friend. She had slipped her collar at Shona Pier on the day of her arrival and was now roaming the island, probably trying to get back to her old home.

'She just hadn't had enough time to get to know me,' he explained. A good shepherd, Iain loved his collies. 'They may not have the same sense towards people – you can't train them to run after a man who's stolen a hundred pounds, like you can an Alsatian, but their intelligence is unique for working sheep. My old bitch doesn't know how to skive like human beings and she works herself to death for me.'

He took a sip from the cup of tea in his calloused hands and went on: 'It's terrible to see her working now. It's all right on the morning gatherings but if we have to go on all day she gets tired and can't run after sheep that are fresh. But she tries until she's ready to drop. She's just too old now, that's why I had to get another.'

He left me a bag of grouse remains and I promised to do my best to catch Queenie.

At dusk two nights later the collie came furtively sniffing between the bracken patches. She ate a piece of grouse but fled tail down when I tried to call her in. That night I rigged up a trap. I threaded light rope through eye-hooks and tied it on to the porch doors, then set a trail of tiny meat scraps to the main bait in the hallway. As soon as I heard Queenie come in I would heave the rope and slam the doors behind her. The first night she did not appear. On the second I went out after two hours – to find all the bait gone. She had sneaked in without a sound!

Next day I had a rough supply trip. Coming back, the south-westerly rain-filled blasts were so strong that even head-on to the waves I was in danger of capsizing. I had to beach the boat in a bay to the south-east because I couldn't take the heavy seas round the last land point. That night, worried about Queenie still out in such cold wet weather, I tied some meat bait with light fishing line to a tobacco tin in my room. As soon as she took the meat the tin would fall loudly and I would slam the doors on her. I sat in the draughty room, eyes aching from the light of a single candle and tired from the hard trip, until 2 a.m., and then fell asleep. Clink – I woke up, reactions slow, and slammed the doors, but Queenie had shot outside again, fast as a cat, and escaped. After two weeks of living wild, she must have forded the low-tide shallows at the island's north-east shore, and had found her way back to her old home in Glen Moidart. Iain collected her by car, kept her indoors for a few days, treating her gently and feeding her well, and she eventually became one of the best sheepdogs he'd ever had.

Sometimes I saw Iain fishing in the big boat way out in the Atlantic. I wondered what he was catching, for with my shellfish baits I seemed able to attract only an occasional pollack less than a foot long. One day at the pier I looked into Iain's boat and was astonished to see a big, wooden box filled with huge pollack, saithe and mackerel he'd caught for the islanders.

'Ye need only the *drogha* (hand line) for the fish here,' he explained. 'And on that you set the darrows, the murderers we call them, because they catch so many!'

He showed me one of his lines on which some two dozen brightly coloured feathers fluttered in the wind. These were the 'darrows', just feathers dyed bright green, red, orange, yellow and white, each tied to a large hook and fixed singly about a foot apart. 'Ye tie a weight to the end of the line, throw it all overboard then jerk it up and down to attract the fish,' said Iain. 'They're irresistible to mackerel who come nearer to the shores and up these lochs in summer. If ye get into a shoal of them it's impossible not to catch all ye need.'

So I bought and made some darrows and one gorgeous late-August afternoon rowed way past the loch mouth a mile or so into the mighty expanse of the Atlantic and began to fish. It was idyllic, sitting there, and with the lion couchant landscape of Eigg, the high mountains of Rhum and the small island of Muck shimmering out on the horizon, it seemed I was almost back among the sea islands of the British Columbian coast. The beauty was intense, blue and gold, enough to make the soul quiver. The blazing colours of the wonderful seascape seemed to drench my retina, and I felt as if I was living in two time phases at once. As I had a perfectly good sea rod left over from Canada I was not using a hand line. Suddenly the rod tip shot under the water and was jerked first one way, then the other. What had I here? I played the fish, its movements quite unlike anything I had ever known before. I had caught two mackerel and was now into a shoal. Every time I threw in I hauled out one, two and even three fish. That was more than enough to keep fresh without a proper fridge.

I rowed past the islets on which the common seals had hauled out to bask in the sun, and they started hunching and sliding into the water as usual, so I cut a few of the fish into three and threw in the pieces. To my delight, although they kept about thirty yards away, the seals began diving and rolling at the surface after the easy food. I whistled and crooned to them and they seemed more curious than before. They popped their heads up and down, snorted, splashed their hind flippers on the surface, submerged and came up somewhere else blinking their dewy eyes, and followed the boat all the way to my beach. As I glided between the marbled islets with their nefarious tribes of gulls, cormorants, ducks, vivid, darting terns and brilliant oystercatchers, the seals splashing about me, it was as if I were indeed in paradise,

After such glorious days I became increasingly reluctant to head south to London, but for the first year and a half I had no other choice. They were long 1,140-mile return drives and I often pulled into little natural dormitories, usually in woods, and slept in the truck.

On each trip there were odd occurrences that made me feel increasingly at odds with modern civilization. Once I forgot to take my cheque book to a bank where I was known and had a small cash-drawing arrangement. No, I couldn't have the usual cheque form, I was told, because the new computer could not process them. So I had to go and fetch the book and walk back again to the bank. I thought computers made life easier! At stores I was asked if I had a credit card. Believing credit buying to be a pernicious influence, making debt honourable and enjoying now what you had not yet *earned*, I had burned the first one sent to me. But I was soon forced into line, and had to get another.

Occasionally I went as a journalist to cocktail receptions but felt bored by the small talk in which little was said, little was meant, and nothing was remembered save those two variations from the truth, compliment or insult. My hankering for the wilds was always there, and I took to hiring boats on the dark Serpentine or Regent's Park lake, where I rowed madly up and down until tired enough to sleep through the traffic din below my little bedsitter. Once I went to Richmond Park to see the fine herd of some seven hundred red deer. Stags with finer antlers than any I had seen on the island, here stood placidly just a few yards away. I met an old man photographing one from a tripod. He told me he'd won a prize that way. No long skilful stalk on the hill – it seemed dishonest for it wasn't really wildlife at all.

My attitudes towards old friends, well placed in public, literary or business life, *had* changed. Yet I sensed an increasing unease, as if some of them secretly felt their existence had become somehow futile, but were unable to take a chance and change their lifestyles. Several times I was told that a simple rural life of meaning no longer existed, or that it was irrelevant. But when you have trekked and slept out in the wild, where animals like grizzlies and cougars roam, nothing is ever quite the same again. The normal goals of 'success', the spare-time enthusiasms of city life such as movies, shows, sports – where most people were merely telly spectators removed from the physical action – now seemed to me colourless and mundane. I felt I had little in common with my own past or the old friends any more, and one by one they all fell away.

Sometimes I came back to the little pier after London visits, having driven through torrential rains in Glencoe, to be told by locals that gales had been raging for days and they hoped my roof was still on. Yet the clouds would miraculously clear as I set out in my boat and I would arrive at the bay of my little paradise in blazing sunshine. Nature's gods seemed to be offering me special privileges, and stage by inexorable stage, the sounds of city life and humanity would die behind me and I would be alone again in paradise. Yet loneliness was always strongest after being in the towns again.

On one London trip I found my bed-sitter-cum-office surrounded by scaffolding and workmen calling to each other all day, so that even writing articles for my keep became impossible. When the popular magazine said they *did* want to run a story about my wilderness life but in interview form, and that they could pay only £200 – take it or leave it – I took it, feeling a man couldn't sink much lower than selling a story about himself.

The end to this uneasy adjustment between my former life and the new came when I returned to my bed-sitter to find it had been burgled. My few belongings – typewriter, camera, two tape-recorders, battery record player, a fan heater – had all gone. I raced round the streets in rage, looking for a 'swarthy man in a brown hat' who had been seen leaving the building with a full plastic sack but, perhaps fortunately, I saw no-one fitting that description.

I felt strongly that these annoyances were more than coincidence, more than fate, almost as if I were being *told* to sever *all* city connections. Galvanized into desperate action, I secured contracts with British and American publishers to write a book about a top movie star whom I had liked and often interviewed. I was determined it would be my last 'patron' from my old life.

I was out of step with current fashions and nihilistic ideas and knew that only nature really interested me. I had grown apart from the mainstream of human life. It saddened but failed to deter me. I had been successful in cities as diverse as Rome, Paris, London and Hollywood; I had travelled widely and enjoyed relationships with interesting people beyond my wildest youthful dreams. Even on the brief trips from the island I had done well to earn a living. But my once cherished goals of success and material independence had proved meaningless for their own sake alone. It was in the wilderness that I must now make my final stand to be a writer, I would finish the movie book, then write only about the things I truly felt. With care I could last out another year. In such determined mood I sold off my former-work files to a magazine, gave up the bed-sitter in London, and drove home.

As if to teach me a lesson for such effrontery, a wild September storm was blowing when I arrived back at the little pier, and with the gales blasting the loch from the west-north-west I had to wait out the night in the Land Rover. As I toiled up the steep slope with a heavy load next morning, the biting midges seemed even worse than in mid-summer. Midges are an economic factor in the Highlands for they hinder work outdoors below a height of 1,800 feet, pester holiday makers and affect movements of wildlife too. Many a time I saw deer restrict their grazing to the windiest parts of the high hills in summer to avoid midges and other biting flies. Unfortunately midges have proved impossible to control for their larvae can live in the tiniest water pools, even in sodden vegetation.

When I walked up the path with the last load, I was rewarded with my first sight of a rare merlin falcon, too small, fast and dark slatey-blue for a sparrowhawk, arrowing like a tiny meteor between the rocks and stunted trees, ready to tackle bird prey as large as itself.

What a treat it was to be back. The Tilley lamp hissed away in the darkness and I arranged my few letters, business ones on top, personal ones below, and read what others had taken the trouble to write to me, sipping a few glasses of wine as my stew bubbled. A letter from an American publisher contained a blow. He had turned down my Canadian book saying, 'There is no market for wilderness books today'. But he added a note that a leading American colour magazine was looking for 'backwoods stories'. I sent off a story I had written about days as a deck hand on the salmon boats along the wild British Columbian coast. When it came out with elaborate artwork weeks later it seemed a breakthrough – the first article I had published about my personal experiences after twenty-one years of writing only of the doings and opinions of others.

By now the contrasts of autumn were showing on the wild seascape, the leaves were swirling from the alders on shore, the bracken was turning yellow, and the first ash leaves were flapping against the croft windows. I went to work again on the cabin, nailing down the whole floor. While moving a large rock beneath it to put in a centre support post, the rock slipped and crushed the abdomen of a coach-horse beetle. As it reared up in pain, jaws champing, feet clawing at dust as it fought for life, my eyes were only inches away and I felt its entire agony. There was nothing to do but end its torture with a swift knuckle. I felt sad. How many creatures do we kill with a single step, without even knowing? But the world was made thus. There is an awful waste in

nature. Many times I saw newly hatched moths, unlucky enough
to emerge on a stormy day, dashed to pieces on rocks by winds,
or pinned upside down by gales and rain, wings smashed, having
made only one brief pointless flight. If we are only part of nature,
I felt, we are doomed.

It often seemed to me that the western Christian intellectual
had become divorced from nature, like most of mankind, preach-
ing homilies about a god of love, seemingly unaware of the harsh
reality in the kingdoms of the wild. Surely it is man's duty to
protect and enhance the last of the natural world. Animals are
innocent. We have creative intelligence and foresight but abuse
our privilege for short-term gain.

By late September I had framed the walls, cut the roof beams
and only the side cladding and roof needed completion. But with
the nights drawing in, the rains increasing, I decided to lay it all
down and cover it for the winter. I would finish the cabin in early
spring.

Coming back up the beach on a cold drizzly day I found two
buff-tailed humble-bees helplessly dying as they clung to the last
purple knapweed flowers. I had never seen bees work harder,
lurching until well into autumn from flower to sparse flower, even
when, five minutes before, I had seen other bees denude the same
flower of all nectar. Now, defeated by the cold and rain, they
were still at work, probing their thick tongues painfully into
flowers in dying slow motion. Such blind tragic dedication, it
seemed, yet if only *we* could work like that, preserve in our own
lives this god-like creative impulse. Was it not the great Goethe
who tried to write a final poem with a finger upon his quilt even
as he died?

As I stacked the tools in the far room and lit the paraffin lamp
a robin began to sing in the ash trees. I wondered if he thought
the light was a new dawn for he flew to the scraps table and began
to feed.

Suddenly there was a loud buzzing in a far corner. A huge
female humble-bee had come in for winter hibernation. She flew

below the roof, examined a small hole, then with a last '*zizz*' of wings disappeared into it. She would survive there until the spring, emerging again on a warm day to find an earth hole, there to raise a new colony. Those I had seen dying were the males, the workers.

Mid-October produced the kind of glorious, eight-day Indian summer which seems unique to the Highlands. From clear skies the sun burned with a fierce heat upon the deep cobalt ocean waters, a total contrast from the limpid greeny blues and hazy mists of summer. Then came three cloudy and ominously calm days, and on 18 October the first real storms of autumn began. As I worked at my desk in late afternoon 70 mph gales swept the land from the west, bending the trees until they seemed to be holding hands over their heads. The wooden walls bulged inwards and every slight gap in the boarding leaked spurts of draught. The place was 'air conditioned' all right. Cold air forced its way up through the floorboards, my feet chilled for the first time, as if the old paraffin heater had not been lit at all. I forgot the gales as best I could – then suddenly remembered the boat. I hastened down with a torch in the darkness, struggling to keep my footing on the steep windy path. The boat was banging about in the rough seas, heightened two feet by the power of the gales, and amid the crashing waves it was now filled with shingle, seaweed and water. I struggled to empty it, to pull it higher, scooping half a bucketful out at a time with fast, flailing strokes of an oar and filling the air with spray, but the sea filled it faster, and in the dark and rain I had to give up.

At dawn, with gales still raging, I hurried to the beach. The tide was down but the boat was now covered in weed, shingle and sand, smashed up against the green grass of the bank like a wreck that had been there for years. With frozen fingers I tore the seaweed from the anchor and bow ropes, bailed the boat empty again, then hauled it higher up the bank. It was holed at last, two jagged rents on the starboard chines, but luckily they were small. I made a rain screen from a sheet of plastic, melted pitch with a candle flame to block the holes, then painted all white again. What a fool I'd been for not buying a fibreglass repair kit. Yet I had been lucky, for if the boat had been smashed up completely it would have been a disaster more significant in my island life than the loss of a car in any mainland city.

6

The Final Break

Never more than during the autumn storms does the island dweller learn that his life is totally controlled by the moods of the sea and its tides. Once, after a dinner invitation to discuss the lease of some land further up the loch, I returned at midnight to find the boat had dragged its rear anchors in the gales. At low tide it was tipped sideways on weed-covered rocks, the engine down and completely submerged in the sea. I shivered through the night without bedding in the Land Rover and when the tide came in again, had to row all the way home in the pre-dawn twilight. I made it only by rowing halfway *with* the winds until in the island's lee; it then took a further two hours to fight the rest of the way against the choppy waves, making only a yard of headway for each hard stroke of the oars. I had to carry the engine up to the croft, dunk it into the fresh water of the burn to prevent salt corrosion of its innards, and spend the day drying it out. That was my first and last 'night out' on the mainland for over two years. The land deal didn't work out either.

Every boat trip was a different battle according to wind direction. If going home, and the gales were from the north-east, I had to slide along in the lee of the Clanranald castle and Riska island, dash across open sea to the Shona shore, huddle along in the lee until another blasting in the exposed area below Arean, avoid the two rocky islets which the seals used as haul-out beds in summer, be shoved at great speed by the waves through the narrow gap between the mainland rocks and a small island, then limp thankfully into my bay.

Going out on a north wind there was a three-way fight: first, to keep ahead of the huge troughs until about half way; second, to negotiate turbulent choppy waters where the two curving streams of wind round the island met; and third, to fight *against* big waves as I reached the main blast of the wind curling round the east edge of the island. In south gales it was hard, sometimes impossible, to launch the boat from my beach at all. At best two strong kicks and thrusts with an oar against the bottom won a few yards but if the engine didn't start with the first two pulls, I had to leap out to prevent the boat being dashed on shore, keep its bow to the waves, and try again.

At least with the south-easterlies it was easier coming home, for one was literally blown back, and I came to call this 'going downhill'. But as November wore on the gales came mostly from the north, and by mid-month it was so cold I had to find a better way of keeping warm while writing at the desk. Some 37 million working days are lost each year in Britain through rheumatism and arthritis alone, and I couldn't afford to lose even one.

Although I had now insulated the double wall boards with bunched up newspapers and set sheets behind the hardboard, the cold winds still penetrated every tiny crack like sharp knives. If I put the old paraffin heater near the window on my left, these shafts were warmer but the fumes were blown directly into my face. If I put it on my right, my left side slowly froze. So I invented a transparent plastic tent and rigged it right round the desk with wooden poles, complete with a roof just two feet above my head. I placed the desk on little wooden blocks so that I could slide the heater a few inches under it. In this way the hot air hit the bottom of the desk, blew round in the space below (thus preventing legs and feet becoming frozen) and then seeped up above the desk too. To stop the fumes coming directly up my nose I tacked a thick towel along the desk's front, arranging it tightly round my waist each time I sat down.

I refrained from buying a costly calor gas heater, and humping more heavy cylinders across the stormy loch in a small boat with the other provisions. When winter finally set in I even stopped using the rear room as a kitchen and moved the two-ring camp cooker on to the shelves behind me for the extra warmth from the cooking flames! At night the paraffin lamp on the desk by the window also filled the little tent with more heat. With only a few small trees nearby (which I was loath to cut down) I had no fire-wood but the old boards torn from the croft, or what I fetched on treks to the forest over a mile away. Wood fires became a treat at weekends only.

Life in the wilds had problems enough, due to the isolation alone, but living frugally also forced little self disciplines on my appetites. I not only rationed wood fires but also playing the radio – a little at lunch and two hours at night, looking forward to certain programmes. Often I could get only Radio 2, and rarely Radio 4 at night, which came through better in stormy weather than in clear. My one other treat from civilization (until I brewed my own) was a few glasses of wine before supper. To reduce temptation I refused to stock up drink, and if gales kept me from boating to the store, this often had to be cut to one or even half a glass of tiny sips a night. 'He died for a bottle of wine in a storm' would have been a shabby epitaph! I also rationed musical concerts on the small battery record player to Saturday nights, or after I had fought home with stores and a handful of mail, the big day of the fortnight. Then I'd have a few wee sips, listen to the music, linger for an hour over the letters, teasing myself with quick readings so I could pore over them fully later, or leave the last page unread, so that there would be something to look forward to the next evening. I began to read more books. To discover that George Orwell wrote his essays in isolation too, on the island of Jura, seemed a small source of strength. So did my discovery of Thoreau's *Walden* where thoughts similar to mine, so well expressed, made me feel less alone. To find I had already spent more than twice as long in the wilds as he had in his cabin in the woods of Concord renewed my resolve to carry on.

I used the shortening daylight hours to pound away on my typewriter, at books instead of magazine articles. For exercise I would go out and dash 400 yards up the hill and back, to get the heart pounding at least once a day. It was probably a foolish recipe for keeping fit.

By late November it was dark until 9 a.m. and dusk again at 4 p.m. Chaffinches and blackbirds had now moved to the easier pickings on the mainland for the winter. Only the friendly little robin stayed around for company. One afternoon, the bird table empty, he had been ticking his silvery notes on the big rock when suddenly he launched himself straight at the window, bounced off it and fluttered weakly back to the rock. I opened it. 'Do you want some food?' I said, feeling ludicrous for talking to a wild bird. It dipped its head up and down expectantly, rocking on its thin knock-kneed legs, waiting for me to throw bread, then it dived upon it before I had withdrawn my hand. It could certainly make its wishes known.

One misty morning, leaving early in calmer weather to shop so that I would still have some daylight left for work, I boated fast for the pier – and almost ran down a huge stag in the sea. It was taking a short cut from the peninsula to the mainland, prob-ably to avoid crossing the little road near the ruined castle. Spurts of breath were blasting sideways from its nostrils as if from a tramp steamer, condensing in the cold air, its eyes showing white with fear as it tried to get away from the boat. When its feet touched bottom, it surged out of the water and galloped from sight. For the first time I missed the loss of my camera, though I had seldom taken it anywhere for I did not think of myself as a photographer. That stag would have made a superbly dramatic picture.

In a brief sunny period on 21 November, I trekked to Shona House for my mail, which I now had sent there with the island's post to force myself into taking exercise on foot. As I walked the rough track between the great oaks and conifers, feeling the enchantment of the still and ancient silences of the forest, I was overtaken by a flock of bullfinches, their snow-white rumps flashing like tiny lights as they dodged through the hazel wands of the under storey. Then some of the rosy-breasted males came back, de-budding twigs and making mournful little '*Tui tui*' peeps, like little sprites of the forest. Suddenly there was a thump at my feet as a large, bright red-brown bird with barred plumage shot upwards. It was a woodcock, and it winged powerfully but silently away, staying low with fast-dodging flight through the densest bushes and trees, as if it knew that way it stood less chance of being shot.

Woodcock have increased since protection during its breeding season, and in Scotland through the planting of new forests. When the woodcock floats high with quivering wings on its 'roding' display and to beat the bounds of its territory in spring, few birds are more beautiful in flight. I laughed to myself as I recalled Noël Coward's remark to a hotel waiter when he thought the bird he was eating rather tough. In his fruity but measured voice he had said: 'The trouble is, young man, there is too much wood and not enough cock!'

Before I reached the house I was overtaken by sleety snow, making a hissing sound as it fell through the tree foliage and hit the leaf-strewn forest floor. Among my mail I saw with delight a letter from a pal in Hollywood, and another from a girl I had met in London and liked. I had invited her to visit. Was she really coming up for Christmas? As I left with the mail and two gallons of paraffin in my pack, I paused for a brief chat with Iain MacLellan. Dusk was now falling and a strong wind had started to blow in the trees. He looked up at the black, leaden sky with practised eyes and suggested I stay in the caravan overnight. But I was anxious to return to the little croft and enjoy my mail alone.

In the forest it was not too bad, but when I came out into the high open hills, the steep stone causeway treacherous now with slippery snow, black snags of granite sticking up here and there, I walked into a totally different world. The rushing wind drove the sleet and snowflakes, grey as ash against the sky, directly into my face in the dark.

I cursed myself for not putting new batteries into my pocket torch. I could not find even the little deer and sheep path but had to deduce where it was from the lie of the snow over the land. The blizzard did not let up for a second, the gale shrieked about my ears and my sweater was soon soaked under the thin, worn-out shower jacket. It was impossible to hurry as I was top-heavy with the pack, and the wind nearly blew me off several small cliffs. When the torch batteries gave out, I went low, floundering in peat bogs and snow drifts. I began to feel I might not make it home. But to stop in that wet cold would have been folly and, using the wind and sounds of the crashing seas as direction indicators, I kept going.

At last I came to the dell where the cottage stood and staggered through the door, my feet soaked, body sweaty yet shivering in the wet clothes. Snow fell from me in icy lumps. It was the worst walk ever on Shona. I lit the lamp, prepared some vegetables and with keen anticipation opened the mail. The letter from my pal was amusing, reminding me of a life that now seemed to have belonged to someone else. But the letter from the girl said she had been to Portugal on holiday, and she had met this young fellow ... Next day the radio told me the same blizzards had caused the deaths of six walkers in the Cairngorms.

In mid-December I went across the stormy loch for a brief trip to London to see the publisher of the movie book and to pick up final oddments (including a television set) left in my former landlady's garage. I bought a small rubber boat at wholesale price, for towing behind and easing the load in my main boat. Also in calm weather it would be lighter and faster, thus saving on petrol bills. I arrived back at the little pier at dusk to find raging squalls of hail raking the sea loch. As I was blowing up the rubber boat, estate foreman Phil Corcoran, who had been helping me try to find my own piece of land, drove by. He said the bad weather had not let up once during my absence. A boat going the short mile to the island had lost its engine to the sea the previous night, and that very morning a cottage near his had lost its roof. I knew I couldn't get across to Ballindona that night. As I unloaded, stacking the heaviest gear under a plastic tarp until morning, I saw him smile at the TV set, for he knew I had no electricity.

'I'll not be having any more need for that,' I said on impulse. 'Here, have a TV set!' And I shoved it into the back of his van.

I spent the night in the truck, buffeted by the winds and noisy hail, trying to stem the drops of condensation plopping on to the plastic sheet over my sleeping bag with strategically placed rags. I felt like heading back south for Christmas. In truth, though, I had nowhere now to go. Nothing saddened me more than for the sake of not being alone, accepting a well-meant invitation and sitting in another man's house, seeing his happy family around him, and feeling a failure in that I did not have all that too.

Before dawn the wind dropped slightly and the hail ceased. I loaded both boats until low in the water and set off through choppy but not dangerous waves. Once again, at the rocky islets the seals used in summer, cormorants flew near, veering off at the last moment to pinion away to the horizon. Several herons rose with gawky leaps, and I knew the last one to be Harry for I saw his broken toe clearly. So the sounds of civilization died behind me, and in blazing sunshine I arrived home once more. As I carried up three heavy boxes of books I felt they were now my best friends. One can really *converse* with books for one reads *considered* thoughts.

It took eleven hard trips to carry all the gear up to the cottage. I had nothing at all left in civilization now. It was the final commitment. The sun shone all the time, and when the last tin was carried in, I went out again into the dying sunlight. The horizon was blue and violet with purple-grey clouds racing in, icing the air. The burn gurgled its pure waters and the city world I had just left seemed as remote as Hades. I laid out all the books for sorting, stacked everything in place and lit a wood fire to dry out the damp croft. Only when all was shipshape, fire crackling, the

lamp hissing its golden light, did the storms break again, the sleet
sliding down the windows as if reluctant to melt. As I felt the icy
stabs of wind through tiny apertures, I realised I had struck the
only patch of sun in three weeks. It was hard not to believe in
some god watching over me, that I was meant to be here. I was
home. I was alone, and I was free.

My euphoria dampened when I found that mice had gnawed a
hole in the door and broken into the food packets above the
cooker. Rice and pearl barley had been stuffed into a shoe and
inside the rolled-up door mat. When I went to air my blankets by
the fire, I found that mice had fouled the centre of my bed. I
nailed a strip of wood to the door to try and stop the little
nuisances.

At night mice pounded noisily round inside the double walls.
In vain I banged a stick on the bedside wall, hoping to drive them
out, but after a short silence they started again. One morning I
found they had come in through a small hole near the chimney
and had stuffed one of my socks with barley – like a peace offering. So I compromised, tipping a small pile of pearl barley in the
porch each night with some boxes and rags for shelter. It worked
– for a time.

Three days before Christmas the sleet and hail were replaced
by rain, then the rain stopped and the wind veered to the south-
west. During this brief lull I went for a walk westwards along the
shore, saw no bird or animal life at all, then began climbing north
over the hills.

As I ascended higher the wind increased once more, making
the climb-walking easier. When I turned east, heading back far
above the croft, the force of the renewed gale was so great that
my cheeks were blown out and I had to fight downwards through
the heather and over the rock faces, actually leaning on the wind
as it tore at my clothing. Rounding one knoll, I saw what looked
like a piece of cardboard bouncing up over the faces towards me.
As it came nearer I realized it was a large sheet of corrugated
iron. It was coming straight for me, and suddenly I felt as you do

when you dream as a child that an aeroplane is crashing and, no matter which way you run, you know it's going to hit you.

I dodged behind a rock and the sheet wafted past where I had been moments before. I forced my way down until I came to the side of the old croft ruin behind my home. It still had two walls standing and some of its roof on. All the scrub oak, birch and alder trees were bending on the hill, as if tossing their heads in rage, and I saw the old ruin was systematically being destroyed by the raging gale. It bulged and then burst, just as if a bomb had exploded inside. Great sheets of its roof iron, eight feet by three, were being ripped off one by one and sent shooting up over the hill. If anyone had been up there in the way now they would have been sliced in half. It was quite terrifying, and I was afraid my own roof would go too.

Then I saw the bricks being shifted and pushed out before my eyes, as if someone had put an invisible hawser round each and was pulling them out one by one. When I stumbled back into the cottage, the door smashed open noisily in the gale and I struggled to push it shut again. I heard a crash and tinkle in the far room. One of the old panes of glass left in the window had split and half of it had smashed on to the floor. The wooden walls of the cottage were bulging inwards, the roof chattering madly. My thought that I should go out and nail half a sheet of plywood over the living-room window, to prevent it being blown in too, was brief. There was no way a man could have held a piece of flat wood steady enough to nail it down in such a tempest. As I stood in the corner away from the window in my little shelter in that wild place, I was most conscious of my own mortality, a man's puniness alone in nature. Yet the reality of it was somehow exhilarating.

As usual I dreaded Christmas Day. When I woke to torrential rain I took refuge from my isolation by turning on the radio. There had been a record spending spree in Britain and holiday flights to the Mediterranean were double the previous year. A Force 9 gale, 'imminent' in my area and nowhere else, was announced. Maybe the roof *will* blow off today!

Then followed one of those interminable pop music programmes with folk requesting DJs they didn't know to 'wish' love to husband, sister, wife, and names and addresses were reeled off as if from a phone directory. Bleating, nasal, reedy or cat-strangle-voiced singers followed one another in dreary succession, each one being called 'great' or 'wonderful' or 'superb' by the hosts who had more than a vested interest in perpetuating the slop. From one came a lyric 'People who need people are the luckiest people in the world'.

Not if they don't have anyone, they're not, I thought bitterly. In fact to me the precise opposite seemed to be the truth. Then for the first time I heard a song from a new rock musical about the life of Christ! The strange line 'Who do you think you are . . .' was bawled out by a crying female voice, presumably referring to Christ himself. Later some church minister was saying such musicals would lead to a revival of Christianity. What they will be is just more linchpins in further reducing the recorded life and example of the man-Christ, the truth he had lived unto death – that only love and creation-in-love redeem – to the level of another mere folk tale. To sheer away the superstitious symbols, props and dogmatic assertions clapped on later by the many often-profiteering churches was one thing, but to replace the brave subtle philosophy with slick sentimentality was no whit better. I turned the radio off for good that day.

In the silence that followed I felt increasingly lonely, but I wrote until near dark, then cooked a usual lunch, adding chips and lighting a few candles as my only concession to Christmas. As the chips crackled in the pan I realised I had been unconsciously preparing myself for the year-end holiday – refraining from drinking for a few days so I would have a little treat in store for Christmas itself. But now, as I sipped sherry and wine before eating, I realised I was acting from mere habit, forcing myself to enjoy in the traditional manner. With two half bottles gone I knew there was no point in drinking alone and put them away.

After lunch I wiped my sweatered arm over my records and put on the battery player my favourite Berlioz, Haydn, Mozart and Beethoven. Had not the great pianist Vladimir Horowitz been a recluse for twelve years too? I was alone, with no loved woman at my side, but as the Eroica boomed through the croft from the wooden box in which I had set the tiny speaker, I recalled that many fine creators had been lonely men, often unsuccessful in their own lifetimes. Beethoven never married. Haydn's great love went off to a convent. Berlioz had disastrous love affairs. Bizet died before 'Carmen' was performed, never knowing of its success. Rembrandt died in poverty. Schubert made a mere £600 from his work during his lifetime. Van Gogh, whose paintings now fetch millions, sold only one in his lifetime (to his brother) and shot himself in despair.

The misunderstood Nietszche and the exquisite nature poet John Clare died in asylums. The rural morons about him called the great Richard Jefferies 'Loony Dick'. Even Thoreau had died a failure at forty-four, having himself had to pay for some of his work to be published. But none of this now seemed to me to be tragedy for I realised that all true creators have a reward which is denied to most – the joy of creation itself. Approval by others is merely the seal upon the treasure chest. It seemed to me the only real truth for an artist of any kind is to smile ironically at misfortune – and to keep trying.

Quite suddenly a memory of the actress Brigitte Bardot came into my mind. We had been talking for hours in her flat in the Avenue Paul Doumer in Paris in the early 1960s when I asked her if, in spite of her numerous romances, she ever felt lonely? She replied at length, and I now recalled the sad and serious expression on that dramatic face as she ended with 'Everybody is really alone. You are always alone. You are alone when you are born and you are alone when you die. All your life you are alone. To me it is normal.'

As I mused a small white moth suddenly leaped from a niche in the woodwork, flew wildly around, then with a brief 'whoff' burnt itself in the flame of the candle, and fell to the table beside me. It stayed there a moment, rearing up on its front legs as if beseeching forgiveness, then it flew up again, hit the flame and sank on to its tail, its wings drawn back like an archangel's, its feet praying against the wick. Its long proboscis glowed red like hot coals and the whole body went dark as the wax spread up it. Then suddenly, with a little pop, it burst into flame and became part of the wick, flames now springing from its head. Poor moth! It could not resist the flame and so it had died before its time.

While the rain sounded softly on the tin roof, almost in harmony with the fine music now filling the little room, I needed to take my mind away from negative musings and self-pity. I selected two new books I had bought by the writer I admired more than most, Henry Williamson. As I picked up the first I felt startled by its title, *The Dream Of Fair Women*. In the book Williamson's mystic, pantheistic, sensitive hero Willie Maddison retires from the horrors of war to a lonely Devon cottage to be at one with nature, and as I read Williamson's prose I found myself transfixed by his clear vision of beauty, of man and nature. For instance, he had Maddison write:

> I can sympathize with all men because I am a man who wants natural happiness, someone to love, and someone to love me, to live with me in my cottage, to guard the well of my spirit where I draw Truth which is also my life. The day after I met you I sat here and loved the sun, the sea and the sky. Suddenly I was afraid; for I can love all these things but they do not respond. I realized that I should grow old, that I should die, and still the wind would shake the poppy, the blue butterfly seek the harebell, and trefoil be yellow on the hillside. I shall be gone, dead, and nothing I can do now can avert that . . .

Even out of context I felt thrilled by passage after passage:

And certain idealists, whom men call fanatics and lunatics and criminals, try to break down the old civilization, hoping to recreate a better world of men but all is foredoomed to failure until the extra wisdom has come into men's minds. They neglect the secret of the woods and fields and how they expand man's spirit if he knows them when little . . .

I turned to the second book, *The Pathway,* realizing that his story of young Maddison was really confession on the part of one of Britain's greatest writers, yet one that has often been written off by effete intellectuals who do not understand what man has lost by severing himself too far from the natural world, who call him a mere 'nature writer'. I turned the pages again, finding more gems for my own heart:

> . . . I realized that all the world was built up of thought; that the ideals which animated the world were but thought, mostly mediocre and selfish thought. Change thought and you change the world.

I read on, surprised such words had been written in the early 'thirties, finding more passages that were so beautiful, so close to my own feelings and unpublished writings, I felt profoundly moved. It was as if Williamson was there with me, talking to me, his words giving me new strength. Suddenly I felt lonely no longer and it seemed there could be nothing more wonderful in the world than to sit there alone, a fire of gathered wood crackling behind me, reading such words. I was no longer unhappy for now I *knew* in my heart, really for the first time, that I was not withdrawing from reality, that loneliness was often the spiritual state of one who wished to create, successful or not. I would finish the commercial book before the old year was over and that would be farewell for ever to that kind of writing. As I rose to prepare supper, an odd thought came to me. In work I trust – all else is therapy.

Boxing Day dawned in a radiantly clear sky. For a while I worked on my book but when the sun streamed through the window after midday, I felt uncomfortable and went to look at the old ruined croft above. The storms had scattered the rotting boards over the ground, providing a fine source of new firewood, and I carried armfuls down to the croft.

I typed, on average, 5,000 words a day for the next four days before running out of carbons. By running the old ones over the top of the paraffin heater, so the remaining ink melted and spread itself over the surfaces, I was able to make ten more copies from each! On New Year's Eve, desperate to finish in time, I typed 14,000 words, a personal record of sheer slogging work. I told myself I was just sitting there and tapping keys after all. It *had* to be done, no-one was going to help, and in the end sheer anger took me through. My stomach and back ached, my fingers felt weak across their backs, but by midnight I had the book beaten.

As the last words went down on paper Big Ben on the radio chimed in the New Year. While my pot of stew simmered away, I started flicking through the pages of a nature book I had bought on my last trip. To my surprise I could identify almost as many indigenous birds and butterflies as I could as a nature-loving boy. My glance fell upon the walnut sparrowhawk I'd lovingly carved at boarding school. Somehow it had stayed with me through all my travels and was now standing on the window shelf. Then I remembered something else and began to search among my boxes of forgotten books until I came to one wrapped carefully in plastic bags. It was a book about birds and butterflies I had written and painted in Sussex at the age of fifteen. My heart surged and for the first time in over 25 years everything seemed to fall into place in my mind.

The pattern had become crystal clear. Nature had been the one constant love throughout life. But it was no longer enough merely to enjoy the wilderness, use it as a retreat or for inspiration. I had to try and *pay it back*. It was then I made my one New Year resolution. From henceforth I would write only about nature and the last wild places and man's place in and influence upon them. I would finally abandon the novel and re-write the Canadian wilderness book until it was the best I could possibly make it. That would be the indoor work while outside I would actively study the natural world in which I now found myself, the Scottish Highlands, one of the last truly wild areas left in Europe. Perhaps after 22 years of journalism, meeting man at his best and worst in half the western world, I could avoid the narrow, specialized naturalist's view. I knew now that only by giving myself totally to this new life, by trying to understand the magnificent Highland wilderness deeply, factually, and writing about it with reverence and love, had I the faintest chance of succeeding.

The decision made, I suddenly remembered my wise old Indian tracker in grizzly country, Pappy Tihoni, and how each New Year he slept out under the open skies in a symbolic act of spiritual renewal with nature. I took a double sleeping bag outside, into a world of violent indigo, the colour of ink, and there beneath the ash trees waving their heads in the wind as if drowning, I slept peacefully until dawn.

7

Winter Furies

The edges of the sleeping bag crackled with the ice-frost from my breath as I climbed out into New Year's Day. The sky was as bright as a metal sheet from the light of the sun which was still skulking behind the hills. Unable to face my desk again, I set off on a trek along the high path to Shoe Bay and back along the shore to take a look at the cabin.

It was good to be out in the sharp, clear air, boots cracking the ice of small pools along the rocky track, the sweet tang of the sea filling my lungs. From the ridge above the blue lagoon I saw an otter leap from a rock, skewer down and turn to swim round a bend out of sight. Damn, I should have gone slower.

On the way back to the cabin, the sun now lighting the sea to amethyst blue, a fish leaped from the mill pond surface and fell back with a noisy splash. A bunch of black winter gnats danced above the high rock, up and down, up and down, as if with glee. Overhead the tiny kestrel swooped down to mob the huge female buzzard which sailed serenely on, taking absolutely no notice. As I crept near the little trees that surrounded the cabin, I was suddenly aware of a tiny head in the crushed brown bracken, like a miniature tank turret with two black beady eyes set high up. Those eyes had seen me too, of course, and now the woodcock winged away like a bronze arrow. This bird can literally see all round itself at any time, and has better binocular vision to the rear than to the front. It was the first time I'd been able to spot a sitting woodcock, before it flew away, since boyhood in the Sussex woods.

As I stood silently amid the leafless trees, noting that the coverings of the cabin floor and framed walls were still in place, the smaller male buzzard (probably following its mate) beat its way west overhead, saw me and immediately veered away with a noisy woofing of wings. It had not expected a human to be standing there.

Then I saw a female eider duck, her thick bandy legs wide apart, her profile Roman-nosed like a fighter's, staring at me from the shore. She sploshed into the water and paddled away. An eider drake in resplendent black-and-white evening dress came swimming along, did not see me, and went as if to peck the duck's tail. She pattered over the surface with fast-beating wings, closely followed by the male. The bright sunny day had perhaps aroused his courtship instincts early. Suddenly a female merganser popped up to the surface with a small, red shore rockling which she had caught in her saw-bill by diving. Next moment the brightly coloured drake came up beside her. With his chestnut breast, white neck band and double spiky crests on his bottle-green head he looked like a duck masquerading as Woody Woodpecker. Just then both birds saw me and began to run across the water. As they became airborne their creamy wing patches flickered like candle flames. Now little brown wrens were flitting around me, sounding their alarm notes like clock springs unwinding in the bracken.

I felt an intruder, and a disquieting thought came to me. What if the woodcock, eider ducks, mergansers or even the little wrens normally nested near the cabin site? Why was I intruding into what had hitherto been a wild area when I had my own croft but half a mile away? Surely, if I cared about the beauty of wild untouched places, it was selfish of me to build in one. I, claiming to be a wildlife lover, also needed to get my attitudes straight! All enthusiasm to complete the cabin waned right then. And when, over lunch at the big house, the Vanes gently hinted there might be problems with the planning authorities, they were surprised when I agreed with them. I decided to leave the area undisturbed during the nesting season, take the cabin to pieces in the summer and use the timber for more repairs to the croft. That was the end of the cabin project.

Such fine days as there had been were now rare. December and January were the most depressing months. With the low sun hidden behind the hills to the south, barely illuminating the landscape; it was dark until 10 a.m. and I had to light the lamp again before 4 p.m. Squalls of sleet and hail were often replaced by rain-filled gales for day after grey day, and now I saw some of the problems wildlife faces in winter.

Harry and other herons were fishing in nearly every sea bay and I would often see him standing for an hour or more, head hunched into shoulders, peering into the water with nothing at all coming within range of his darting beak. Just how a hard winter can decimate heron numbers was shown by the icy spell of 1947 when nearly half Britain's heron population perished. They have even been found with water frozen round their legs. After the long cold winter of 1962/63, when more snow fell in Britain than at any time in the previous 150 years, occupied nests declined from 3,700 to 2,200.

I discovered that red deer now had to forage over a wider range. In summer, when growth was lush, they kept mostly to the cool high tops, away from midges, tabanids and other fly pests, only coming lower at dusk. But now they were venturing close to the croft for mosses, lichens, heather, dead nettles, grass and to clip the tops off young bushes and trees, or to eat all the bramble leaves they could find. Brambles keep many of their leaves until well into January and sprout them again before most bushes. I woke once to find a young stag cropping the leaves from the blackberry bushes that lined parts of the burn.

Sheep also eat bramble leaves but twice that winter I found these bushes had drawbacks for these animals. After heavy rains I went down the path on 8 January to find a large ewe caught by the incurved thorns on two thick bramble leaders. She was sitting down on a rock in the raging icy water of the burn, her head lolling weakly, stoically awaiting death by cold and starvation or perhaps drowning once her front legs could no longer support her. The ground was splashed along the banks where she had tried to break free.

As I stepped into the pool to pull the brambles from her wool she struggled. My hands were soon gashed and bleeding, but bit by bit I freed her, put my hands under her soft bloated belly and lifted her up on to the grass. She could hardly totter away, and after a few yards she began to graze weakly. A thick woolly coat may look warm to us but it clearly has its drawbacks when you have only a domestic sheep's brain.

Five days later I started on a trek up the hills behind the croft where I found an older ewe, caught up on a briar out in the open. This one had stumbled in constant circles, tying the thick bramble tighter round her neck until she stood unable to move, her head bowed, awaiting death. This time, remembering my gashed hands, I went back for gloves and a knife to cut her free. How stupid they seemed, for all they had to do was to gnaw through the brambles. I recalled the wild sheep I'd seen in Canada, the mighty bighorn, now rare, whose rams could hurl a wolf from a mountain ledge. These man-bred creatures seemed altogether different.

The battle with the noisy woodmice was still a nightly exercise, for my softhearted porch feeding had merely increased their numbers. One night I was woken by a rustling noise. When I shone the torch I found a shrew beside the cooker, chewing at greaseproof paper on which I had left some batter mixture and butter. What a racket! It froze for a second but as I got out of bed it leaped three feet to the ground and vanished into a tiny hole by the fireplace. Until then I had always believed the shrew, our smallest mammal, to be a complete insectivore, feeding on beetles, larvae and worms. It was interesting to discover it would eat such kitchen food. Later I found that shrews will feed from the fresh carrion of deer carcasses.

Normally I wrote only on the stormy days, but I became so engrossed in the Canadian wilderness book that I hated to lose momentum. After a two-week spell at this (my movie book unposted) I had to go out for stores when the westerly gales were tearing off the wave tops. I managed to ride them 'downhill' and left the boat in the lee of the ruined castle because the sea round the pier was raging against the sea wall. On the way back I had a real battle, boat banging in the troughs, almost capsizing, and water spewing over the bows. The sleet was blinding, my back and neck soaked in icy spray, my free hand clutched below my knee for warmth. I felt a sudden rage at the storm for making an already hard life more difficult. I cursed the waves aloud – 'Damn you, I'm *going* to get there and you won't stop me!' – as I struggled to keep the boat on course and actually stay in it. The gales had piled huge banks of slippery weed on shore, and when I reached the bay at high tide I just hauled the boat on to it, without the protecting cradle.

I felt victorious, all loneliness gone. What a pleasure it was to reach the little sanctuary, knowing I had posted off the movie book, a year's work, the end of all commercial writing. I prepared my supper, sang noisily my favourite musical roles, banged about at will, held loud, imaginary, victorious arguments with those who constantly turned down my writings – all without fear of disturbing my neighbours, human or animal. While silence is important for a man trying to live close to wildlife I realised there are plenty of noises in nature too – stags roar, owls hoot, seals bark, gales howl, and thunder imposes its awesome tympani upon the hills and glens. As long as I kept my little celebration inside four walls I was doing no harm. What, after all, were the important things – doing the work one really loves, reading a good book, eating a simple meal, saving one's boat from a storm, one's life from the sea, animals from death, reaching out to contact others, if only in letters or books. How meaningful life suddenly seemed. I felt happy again. Loneliness in the wilds, if only for a spell, should be compulsory for everyone!

Later I read my mail while listening to Bach's magnificent Toccata and Fugue in D Minor. I had no need to imagine the cathedral, made by man to the glory of God, in which it was played. Was I not hearing it in God's *own* cathedral, the wilderness, the natural world built *by* God?

Even the friendly robin seemed to have gone and now my only pal was a large spider I called Sarah who had made a web on the window in the west room. Patiently she sat in a corner, seemingly able to go without food for weeks. After a sunny day, when a large fly buzzed into the lamp and dropped to the desk with singed wings, I put it into her web. She came down in a flash, wrapped it in silk, sucked all the juices, then two hours later cut the remains free and retired to her top corner without bothering to repair her web first. If she woke up hungry she now had less chance of a new meal. A week later I took pity on her again and put a lighted candle near her web. The few flies that were roused by brief sunny periods flew towards the flame at night and some caught in her web. Thus she got her meals with no effort from me.

By mid-January I had a way of taking hard and useful exercise to offset the spreading waist that afflicts the writer: I started a garden, clearing over a ton of rocks from a ten-foot-by-eight-foot patch below my window, carrying up sacks of seaweed from the beach to fertilize the ground for spring planting. I took the fine earth from mole hills over a wide area to replace the rocks with useful soil.

One day, with slight snow falling, I took a long walk along the high track to Shona House for my mail, returning along the lower path through the wild enchanted forest. The snow could not settle, and I felt I was at Camelot where it always blew away – at night, of course! I walked up a lovely green dell which the islanders called 'Lovers' Walk' though it had been a long time since any young lovers had been up there. Nor did any fine lassie emerge from the trees to play Guenevere to my jaded Lancelot. But deer trails there were aplenty, the deep twin slots showing in muddy patches, and a flock of woodpigeons had moved in to colonize new roosts in the conifers. I had seldom been in that wood without seeing something new and that day was no exception.

Suddenly a huge bird, deep russet in colour, went up about thirty yards ahead and, with rapid wingbeats, interspersed with short glides with its wings curved downwards like hooks, disappeared between the trees. It went too fast for me to get the field glass on to it. It was too large to be a grouse, more like a small turkey, and for so heavy a bird it flew with unusual silence. It had to be a female capercaillie, an exciting sight for it was rare in the western Highlands.

There were so many windfalls in the forest that I had to keep crossing a deep burn gorge. Being less common after years of felling in the Highlands, such woods are precious – jewels among the hills to be preserved at any cost.

I left the gorge and emerged on a small woodland plateau with the flat green surfaces covered in moss of an ancient rockfall just below me. Suddenly there was a flurry, a scuttering sound, and a large tawny animal with pointed ears and a thick black-banded tail dived off one of the rocks and disappeared between them. Again I had no time to get out my field glass. Had I been in Canada I would have sworn it was a raccoon but here it just had to be a big cat, almost certainly a wildcat, and this was only my second glimpse of one. I went below the rocks, searching for tracks, and there in the mud I found two unmistakable four-toed prints without traces of the claws.

After leaving the forest on the way home, I saw the kestrel being blown like a matchstick over the hills, its wings sharp-pointed in the wind, so that it looked like a tiny cross against the sky. I gave a high piercing '*wheeoo*' whistle and it actually turned round in the air and came towards me for a few yards, its wings bending supply, before reverting to its original course. Now I could use the field glass and watch it flying and hovering, its wings making extraordinary flickering adjustments to the winds. Suddenly it went down, hovered, lower, hovered again, then it dropped. It flew up in a few seconds with something small and black in its talons, probably a beetle. It must have had a hard time hunting with snow still falling.

I tried setting out night lines with a dozen hooks on each baited with bacon scraps and mussels, but it seemed that most of the fish had deserted the winter cold of the shores for the deeper, warmer waters of the Atlantic. Apart from an occasional ten-inch saithe, I caught nothing at all. The rough seas frayed and drifted loose weed against my lines. Shellfish too were rarer, and in one long search all I could find were a few winkles and *one* cockle under some seaweed. Even the mussels seemed to have halved their populations, and I discovered many had half buried themselves in the gravel and sand, as if seeking a more stable bed in the wind-blown tides. Although I searched through abstruse biological literature about mussels on a later trip south, I found nothing to explain this phenomenon.

By late January flocks of curlews, flying with oddly slow wing-beats, were working the tidal sand flats near the pier and castle. They can use their uniquely long curved beaks to probe for worms, molluscs and other small marine fauna deeper than any other wader finds possible. Often I just drifted in the boat to hear their haunting '*kourlee*' calls, the musical embodiment of the melancholy winter scene. Occasionally small groups of winter-visiting golden eye ducks from northern Europe came over the boat, wings whistling creakily from their squat bodies. They glided down to the river mouth, stretching out their ludicrous webbed feet, like air brakes, before landing in the water. Then the white-breasted, black-backed drakes swam fussily round the browner females, flashing their circular white cheek patches as if they were silver dollars for the ladies to take.

Although snow had fallen heavily in the last few days, the westerly winds off the ocean kept the island ground just above freezing point, and it melted again each day. In a rare sunny interval on 8 February, which made me feel that maybe spring was on the way, I saw the male buzzard, a feather missing from its right wing, flying over the shore, being plagued again by the cheeky kestrel. The small falcon swooped down past it from above, then looped up again from below like a yo-yo on invisible strings, as if showing off its greater agility in the air. Too small and light-coloured for a peregrine, it was moving exceptionally fast for a kestrel, and it clearly regarded the bigger buzzard as a poacher on its territory. The buzzard put on speed, to escape its tiny tormentor, looking like a dignified schoolmaster forced to put up with the shenanigans of a cocky student. I had been making notes on both birds' movements and it seemed the buzzard worked the area once every four days while the kestrel, covering a much smaller territory, was hunting over it every two.

That brief, sunny period proved a mere glimpse of spring because overnight the wind changed to the north and hail froze to the grass and plant stems and in tiny hillocks along my wooden sills. Even the water in the buckets inside the room froze on top. Because my boat engine was now often misfiring I brought it up to the croft, but my hand melted the ice on its handle and it slipped from my grasp and fell, smashing its lid upon a rock. I repaired it as best I could with wire and cleaned the points and plugs.

To post urgent mail on 12 February, I took the boat into the teeth of the hailstorm that had swept the land for three days. Hoping the engine wouldn't conk out, I kept to the lee of every rock and islet for a brief bucking pause before dashing across the main water. The skies were grey and violet, the winds most violent when they preceded the hail clouds that looked like great jellyfish dangling grey curtains of stingers. I could see the squalls coming. Wave tips a few hundred yards away would be torn off first, then the hail and dark was upon me, the boat bucking like a bronco as I tried to keep it straight, the stones – some half an inch across – rattling, bouncing and hissing down upon and around me. Then the squall would pass and in a brief glimpse of blue sky and blinding light the sea would turn a frothing bright green and white. Coming back was worse, and I had to shelter in a small horseshoe bay below Shona forest and wait until the next brief lull. At times I had to slide rapidly into the centre of the boat, my arm stretched out to the throttle, to keep my weight off the stern in the deep following waves. I don't mind admitting I prayed a little on such trips. This was no country for atheists.

It must have been instinct that made me go out that day, despite the gales, for I woke next morning to a wild blizzard and a world of white. Sleet among the snow came down the chimney and tickled the paper in the fireplace, like mice rustling for food. Vortexes of air in miniature whirlwinds tore in from nowhere, blasting between the high rocks and across the ash trees, tearing off the last old, brown leaves from the scrub oaks and sending them soaring into the sky. The gusts ripped at the roof so that the iron sheets chattered like false teeth, the gutterings rattled and the doors creaked, swelling inwards slightly like the windward wooden walls themselves. Sheep wandered, staring perplexed at a land they had not seen before, going '*mairr mairr*' in bewilderment and frustration. The sea mashed angrily below, white percherons romping against the rocky teeth and scrabbling over the granite, up, back and down into itself again. The great boomings and roarings of the surf were followed by smashings and seethings and the clacking grind of stones. All around me the mountains crouched and hid in their blinding cocoons of snow.

Yet later in the morning small white clouds, like fleece, scudded across a deep blue sky and the sun burned with a fierce primeval heat. I sat in the porchway, sheltered from the northwest winds, a fish box for a chair, stripped to the waist with sweaters tucked round the cold parts in shadow, and roasted with the heat. Minutes later it was so dark I could not see the beach or my boat and the storm was back, driving the flakes of snow against a grey-green sky. Looking up into that sky now was like looking into the jaws of hell, with the smoky smuts from the fires coming into my eyes, and the hail so bad it smashed through my eyelashes. I went indoors, my whole face smarting and glowing rosily after the natural hot-cold sauna and skin massage no city masseur could have given.

Snow fell again all night, and after the winds switched round the clock, from north-west to south-west to south-east and to north-east, they lessened to almost total calm. Now the waves that had fallen ferociously, dying on my beach the day before, came in caressingly to fall with soft chuckles in the sunlight between the snow showers. I went for a walk along the south shore into a magnificent tableau. The birch and alder groves were now a fairyland, with every branch, twig and even delicate bud carrying two or three inches of snow, now falling softly without wind, drifting down in a white and silent world. Sheep stood about disconsolately, with one large ram appearing completely baffled. Some were trying to find food, thrusting their heads into the middle of rush clumps and bent grass tussocks. I cleared some open patches with a shovel, only to find the grass and ground now frozen hard underneath. The sheep took no notice of my handiwork.

Before lunch I stumbled through two-foot drifts to the west-ward shores. I was just rounding a small rocky knoll when I saw ahead a young red deer stag with six-point antlers. It was scraping snow from heather tufts with a forefoot. The wind was in the wrong direction, from behind me, and it was too late to go back and in again from a different angle. It got my scent in seconds, looked round briefly, then bounded away, not as light on its feet as usual because of the drifts. Just then the big female buzzard flew from a tree where she had been sitting motionless. She had been watching the deer.

In the afternoon I boated out for supplies. Although the engine now seemed better, I rowed back against slight winds, sticking at it, for one never knows when one will need rowing fitness in a wilderness life. I picked up four fish boxes that had been washed ashore by the gales – always useful for furniture.

A lone black-throated diver, known as 'loons' in North America, appeared to be the only other moving life in the silent air. It flew overhead towards a freshwater loch on the mainland. It seemed a highly competent flier, its pointed wings beating like twanged steel. There was no wastage about it, like a snake flying, very sharp everywhere, a beautiful flying and diving machine.

That day some of the locals told me we were having the worst winter for years, and I was glad to be there to enjoy it, for somehow it made one feel closer to the vital animal state. At dusk all the hills looked different, white while the sky was black, the reverse of the normal, like living in a photo negative, and before the dark finally fell the sea shone with a beautiful turquoise colour.

8

Sheep in Nature's Cycle

Before the end of February many small birds were coming to the bird table. Little Fat Sergeant, the colourful, cock chaffinch with three white stripes on his wing, was back with his demure, grey-green hen, waking me each morning with loud '*pink pinks*'. Three blackbirds had arrived, a cock with black plumage and bright yellow beak, and two brown hen birds which announced their presence by worried '*kip kips*' as they approached the table in fits and starts. A hedge-sparrow also came flitting through the old bracken stubs and threw its bread to the ground where it felt more secure. Missel-thrushes were now singing their stormcock dawn songs in the highest trees, one atop a little spruce that reared its tip against the sky because it was rooted on a high ledge behind the croft.

By the end of the first week in March the daffodils had grown to ten inches, ready to flower. In the grottos along the south shore burn pussy willows were silvering the bushes with shimmering light. Each morning down on the shore brilliant oystercatchers bathed at low tide, flinging water everywhere with their black-and-white wings and orange beaks, preferring where the fresh waters of the burn mingled with the sea. A pair of pied wagtails came in from the east, calling '*tisit tisit*' with each loop of their undulating flight, like a pair of noisy flying scissors. They flitted about the croft with tilting tails. In glorious sunshine on 8 March, my paraffin heater off for the first time, I was startled on my way down to the beach when the first queen humble-bee of the year tried to go down my ear as she searched for nest sites, suddenly filling my world with a loud buzzing sound. Three days later, as I boated out in a dead calm, a great spotted woodpecker drummed loudly on an old tree snag in the Shona woods, and there were small mating flocks of eiders on the sea, the males now sirening their mates with soft '*awhoos*', a sweet music to usher in the warmer days of spring.

But on 12 March there was trouble on the bird table. A new cock chaffinch arrived, his upper wing chevron blazing white and thick, with a new hen who was fatter than the first one. Now he and Little Fat Sergeant were engaged in fights, flying straight up into the air, clawing and pecking at each other so hard that the feathers flew. The new cock seemed to win most of them and to be in charge of the table. Of Sergeant's little mate there now seemed no sign, and I wondered if she had perished in a hail-storm two days earlier.

At dusk now seven hinds, two yearlings and the young stag came behind the croft, so close I could hear them cropping the green sward. There is a theory in the Highlands that old hinds and stags *know* when the stalking season is over – for hinds it ends on 16 February and they then come closer to habitations. One even-ing, the wind from the south-east, I moved silently, a few inches at a time, straight towards them in a straight stalk from the west corner of the croft. I was within eight yards of the nearest before she saw me, then she barked loudly like the '*hoff*' of a black bear, and they all took off. When next morning I saw one of the oyster-catchers standing atop his mate while she held herself prone, I began to wish my camera had not been stolen in the burglary.

By 15 March the first primroses were out in warm, sheltered nooks along the burn but such emerging beauty had no peaceful effect on the chaffinches for they were still scrapping each day. I saw Little Fat Sergeant make a determined attack on the younger cock, driving it to the ground where he won the clawing, pecking brawl, for the new cock suddenly flew away. Sergeant, back in charge of the table again, also chased the new female off the bread. Then he began to peep joyfully.

Three days later, the daffodils now breaking out their yellow petals, I heard a loud '*pink pink*', then the full chipping song of the courting chaffinch. I looked out. Sergeant was in the centre of the table and with him was his slimmer demure mate. He was in fine voice for he kept '*pink*'ing with his mouth stuffed full of crumbs, his little throat bulging to express joy at victory, his wife's return, but it sounded from far away, as if from a ventriloquist. There was now no sign of the other pair.

Sometimes, when I went out with scraps in the early morning, Sergeant flew, with his mate behind, in great looping swoops up the hill from the shore trees a quarter mile away. He was now in full song about the croft, vying for first place on the springtime stage with the thrushes, willow-warblers, wrens and robins.

One day, as I dug and sieved the garden soil, the seaweed rotting down nicely, I saw the kestrel over the croft. I now carried the field glass nearly everywhere and I trained it on the bird. It was hanging in the sky like a fulcrum of flashing copper light. Its small head had amazing mobility, looking round like a capuchin monkey's as it hunted by sight, glaring down at the earth. Not much would miss those eyes.

By the third week in March Charlie Crow and a mate were back at their old nest in the birch tree down by the shore, sheep were '*baah*'ing noisily round the croft in the early hours, the lambing season about to start. I was looking at the daffodils when both hooded crows suddenly winged away shorewards. I had just counted the flowers – fifty-three, and four more than last year – when shepherd Iain MacLellan appeared, checking for new-born lambs. Busy treks on the Hill (as Highlanders call mountains in the plural) now lay ahead for him.

On these rare, early-spring visits I came to know Iain better. He said the island had been his idea of a Highland paradise when he was a youngster, with a crofter in every cottage, each with a couple of cows and about six sheep. They cut hay in the hills, peat for winter fires from the high bogs, ran small vegetable gardens and grew potatoes in fertile patches which they dug by hand into undulating 'lazy beds'. But, even then, the rising costs of food and distribution, the depopulation of the region as more people moved to better-paid jobs in the industrial cities, were changing the old ways. Most crofters then also worked as employees on the estates for the lairds.

'We were taught to serve our masters well and not to waste his money,' Iain said. 'We learned about God and moral life at our mother's knee and that stayed with us through all our lives.' His father James MacLellan had built the fine mahogany desk on which I now worked. 'I used to walk here every Saturday morning, pick up stores left on the beach by the big boat, carry them up to the old croft behind yours, and then sit listening to the old stories of my father and uncle. Life was simple but I'm sure people were happier then. There was no television and we did not crave what others had or a richer life for we hardly knew there was one.'

I showed him how the old ruin behind had finally blown down that winter, wondering if he was not pleased that I was there, but he said 'Ay, 'tis gone now. And this would probably have gone too if ye'd not put the new front on.' He also liked my garden and promised me some fencing. 'One day ye may wake and find deer have had all your vegetables. Ye'll need a high fence and to roof it over.'

Apart from running the island's boat and other estate work, Iain and his co-worker looked after the sheep flocks. It was hard work, far removed from the popular notion of shepherding. There were six gatherings a year when he, his wife Morag, and two or three other men, if they could get them, made several walking swathes over the hard hills to herd the sheep into the fanks on the east side of the island: in February for oral dosing against fluke and worm; in March, ten days before lambing, to dip them against ticks, lice, flies and other insect pests; in June for shearing the yelds, castrating male lambs and marking the sheep; in July for shearing the milk ewes; late August or September for selling surplus lambs, castrated males, some tups (rams) and the old ewes up to ten or twelve years old that probably would not survive the next winter, for mutton; and in October for winter dipping.

I had already found out from the Vanes that then it cost £700 for one man's yearly wages, yet the sheep only brought in some £800 a year, which included hefty government subsidies. These, in Shona's case, were paid on about 250 breeding ewes and in the early 1970s varied from £2 to £3.50 each ewe per year. The island also kept from forty to sixty ewe lambs plus eight to twelve tups for breeding. The truth is that sheep cannot be raised successfully in the Highlands today without these subsidies, which are paid by all British taxpayers – who, of course, benefit from the wool and the meat.

At first it seemed strange to me, an incoming layman, that there should be such a deep attachment to sheep here. It was precisely the coming of the sheep that caused the notorious Highland Clearances between 1782 and 1854, when many Highland lairds sold off their land to incoming southern sheep farmers, as well as for gunsports, and thousands of crofters were thrown out. This indigenous population often had their homes burned down and were compelled to move to industrial towns for work or to subsistence areas on the coast. Many emigrated to Canada and New Zealand.

Certainly I could understand that sheep farming was an effective way to utilize such harsh hill grazings, but it seemed an odd sort of economy if this system could be maintained only with heavy subsidies. The Vanes gave me several reasons for keeping sheep, the least of which was profit. They were now a traditional industry, they caused the employment of a second man as well as occasional helpers, so helping the area's unemployment problem and they prevented the island from becoming too overgrown. Another reason was that Iain himself was a keen shepherd. While he was boatman, ran the island's little saw mill, large vegetable garden and poultry, stalked the deer and tackled most jobs as handyman-carpenter, he liked nothing more than to be on the Hill tending his sheep. Thus, if the sheep had gone, the Vanes might well have lost an essential, possibly irreplaceable, worker too.

I wondered if Iain thought I felt that wildlife suffered from the grazings of sheep. I had as yet no real opinion, though after finding several dead sheep on the island hills in winter, it was obvious such carcases provide valuable winter carrion for golden eagles, buzzards, ravens, crows, wildcats, gulls and even foxes. As far as grazing competition went, red deer had been increasing in recent years in the Highlands.

Iain disclosed one day that his co-worker wanted to leave the island. 'We have a hard task keeping workers here,' he said. 'We've had fourteen in the last ten years, mostly married men. I get tired of ferrying them and all their furniture over and back. They don't like the isolation – or their wives don't.'

Isolation? I thought. What would any married man or woman with other families near by know about true isolation! With a loving wife who felt as I did about the wilderness, I could imagine no finer place to be. But I could see that, if there were problems in a relationship, even such comparative isolation would increase them. I knew many city working couples who survived together because they only took each other in short evening doses.

Then Iain told me of his problems at lambing time, of young lambs falling into crevices, ewes with difficult births, and foxes taking the lambs.

'Often a ewe walks miles from her usual grazing area to give birth. I once found one from this area at Shoe Bay. The pain had made her move and seek cover – I found her dead with her front half wedged under a rock. The lamb was out but dead. Yet the previous morning after an eight-mile walk I'd seen nothing at that spot.

'Once I was up on the highest hill, Beinn a Brailidhe, and saw a ewe lying down. She got up and dashed away fast, then went round in circles. Something was very wrong, as I could see a lamb hanging out. One foot was out below the lamb's chin, as both should be, but the other was bent back behind its head. And when the pain was really bad the ewe charged off like a bullet. I ran down, caught her with my collies and, holding her between my knees, pushed the lamb back and got the other leg out with my fingers. Then she gave birth normally.'

Just then one of the hooded crows flew above the window. I remarked they had been back at their old nest.

'Ay,' said Iain, 'and the best thing to do with it is put a charge of shot right through it!'

I had heard of hoodies preying on sick lambs but I asked if he'd not leave even *one* egg in the nest?

'No, Mr Tomkies, I would not. Hoodies – and ravens – will attack a ewe that's having trouble with her lamb. She tries to stand but because of the pressure and pain she sometimes topples on to her side or back. And she can't get up again if she's in a little dip in the ground. Then the hoodies and ravens come and peck out her eyes, at her tongue, and kill the lamb.'

'Yet animals evolved this way,' I said. 'It's part of nature's system, created by God.'

Iain, a religious man, deigned to answer the blasphemous doubt, that God had perhaps created an imperfect system. Such reasoning was hardly for man. His job was to protect his flocks as best he could.

'I came over to Shoe Bay once,' he continued, 'and I found a ewe trying to give birth, but she had rolled into a hole and couldn't get up. I found her because of the big black-backed gulls, hoodies, ravens and a buzzard circling over her. When I got there she had both eyes out yet was still alive and giving birth. I had no gun with me so I ran all the way home for a knife and back again. It's the best death – as the blood comes out she feels nothing, just fades away until her heart stops.'

Iain reckoned the hoodies had flown off because they saw him coming. They knew *him* as an enemy. My befriending of Charlie had probably made them know I was harmless. He felt hoodies could tell the difference between individual human beings.

'Two years ago, when it was my week on the lambing, my partner was working the garden near the hen run – though he had nothing to do with the hens himself. He saw this hoodie coming and going but paid no attention as it wasn't in the run itself. Now, when hens get broody, they like to have their nests wild, where no-one can find them. What was happening was one of the hens was getting through the wire and laying her eggs near a stream a good way from the run. And that crow was coming to the tree directly above the hen while she was laying. As soon as the hen left the nest to go back to the run for food, down the hoodie came and took the egg. This went on for a whole week. But when I was in the garden the next week, the crow knew me as a personal enemy and never showed up once!'

Iain had work to do and set down his empty tea cup. As he left I said I'd do what I could to check the lambs in my area. While I understood his viewpoint, after returning from North America where the predators were still wolves, coyotes, lynx and occasional cougars and bears, little mention had ever been made of smaller varmints like foxes and crows. It seemed ironic that in Britain, where predatory man had long since exterminated such large carnivores, his focus had now come down to the remaining smaller foxes, stoats and crows.

The handsome hoodie is a relatively rare bird on a world-wide scale, far more so than the raven which *is* protected in Britain (except in Argyll and Skye). The hoodie survives (mainly as winter migrants) only in small numbers in England, chiefly on the north-east coast, and does not exist in Wales. While fairly common in Ireland and with a few in Ayrshire, its main strongholds today are in the north-west Highlands, Europe and Scandinavia. In Russia it is limited to the western edge. Small wonder perhaps that the hoodie, known in Gaelic as 'An t-eun g'un duileachan' (the bird without compassion) has evolved such great cunning to outwit his main enemy.

The day after Iain's visit I had some hilarious entertainment from the bird table. Little Fat Sergeant came back with *two* hens, his own and the fatter one that had been with the other cock, who had not reappeared. As I watched from the window Sergeant's wife chased the other hen off the bread, then pursued her into the bushes for good measure. This battle between the hens was repeated several times while Sergeant went on nonchalantly stuffing himself with crumbs.

The first of April certainly made fools of some flies which had hatched out in the far room. It was a cold day with heavy hailstorms and they walked about bedraggled, wings closed limply over their backs, unable to fly in the new hostile world into which they had emerged. A day later, after a period of sun, the year's first emperor moth appeared on my window, drawn by the effusion of light rather than Rachmaninoff's Second Piano Concerto on my battery record player.

I was working hard on the Canadian book on 6 April when Iain appeared for a quick cup of tea. They were gathering in the sheep before the local foxhounds came on the morrow to try and clear out the foxes before lambing. We talked about foxes, and I said I felt it was wrong to hunt as a sport, to gain pleasure from 'the chase' and the death of a far-outnumbered animal.

'We don't get pleasure,' he explained. 'We're no sitting dainty on horseback. No horse could hunt these hills. We're on foot and it's hard work. I don't like killing anything really but foxes kill many lambs. I know they're not evil. They have their instincts, and you have to pit your instincts against theirs, because they undermine your livelihood.

'From the first, you look after your sheep the best you can. After the first lamb comes you have your hardest time. You have to try and visit them every day, cover the whole island. If you have a hundred ewes you may have a crop of eighty lambs. Then one day you walk the hills and find maybe six lambs, their heads torn off and the rest left to rot. It's waste. When wanton killing occurs like that you just have to go after the foxes.

'When foxes have cubs they often slit lambs and just take the stomach, which is full of its mother's milk, home for the cubs and leave the rest. Later they may kill and just take the tail, ears or head for the growing cubs to play with. I once found nineteen dead lambs in and around a fox's den – a terrible sight.'

He had had many experiences of the fox's cunning. On one hunt with den terriers the vixen had come out and deliberately run close between two men before vanishing behind the rocks, as if knowing that way no-one would dare shoot! When in 1969 an Act was passed banning the gin trap in Scotland, a four-year gap was left before it became law for the invention of a more humane trap to control foxes. One such was the 'drowning set'. It is set in a pool, at least fifteen inches deep, on a small peninsula from which a chunk has been removed, forming an islet. The bait is buried, so making it more selective for foxes. When the fox is caught, it struggles, leaps, and is quickly drowned by the weight of the trap which is tied to a heavy underwater stone. Such pools and peninsulas are not easy to find. Iain made such a pool once – and the fox walked round it, then dug a canal with its paws, so draining the pool!

I heard the hounds baying in the hills next day but all I saw of the hunt was a sudden, whitish shape outside my window. It was an old bitch hound, ribs sticking through her coat, which had broken away from the pack. She put her paws up, calmly stole the bread from the bird table, then sneaked off down to the south shore. It seemed she was not keen on hunting foxes any more and was just looking for food and a quiet life. All that day the two hoodies were absent from my bay. Just before dusk Iain's co-worker stood outside the croft, his shotgun upside down on his foot, and flourished a blind, dead fox cub, blackish-brown, as they are at that age, with great pride. The terriers had gone down one of the dens and had brought out two cubs. The vixen, no doubt trying to stay alive so she could raise whatever cubs might be left, had escaped.

At the time it appeared to me cruel, and that much fuss, time and manpower were being expended for fairly minimal results. I wondered too if the relatively few foxes killed this way were mainly old, diseased and less wary than the young, so perhaps even *helping* to maintain vital fox-breeding populations by weeding out the less fit. Such killing was sporadic and it seemed likely other surplus foxes would come in from outlying areas to take the place of those killed anyway. But I was not a sheep farmer relying on it for a living and I could understand the feeling that *something* had to be done. It just seemed a lot of trouble for an industry that could not survive without subsidies. Indeed, the foxhounds themselves were also subsidized at the time by a £1,000-a-year grant from the Highlands and Islands Development Board.

It is general policy in the Highlands to try to control foxes especially when they are feeding cubs, though it is not certain how many lambs taken are already dead from various causes, including never getting a suck of their mother's milk. Foxes can't eat lambs all the year round but the war against them continues all the year round. Yet they eat many voles which not only compete with sheep and deer for green grass but also act as hosts for a tick which transmits the serious 'louping ill' disease to sheep. Foxes, weasels, stoats, owls and wildcats, against which gamekeepers and some farmers have waged war for years, keep vole numbers down. Foxes especially eat large numbers. In one Highland study by Dr James Lockie and W. N. Charles it was shown that in winter foxes, owls and weasels between them reduced voles from 90 to 45 per acre.

Next morning Charlie Crow and his mate were back on my beach and gave me an exhibition of how they feed on shellfish. One hoodie suddenly flew up in the air with a cockle and dropped it on to the largest rock on the shore. The other bird flew in but the first dived down faster, picked it up, flew further, dropped it again and was down on to it and away in a flash with the succulent titbit from the smashed shell. I wondered if crows had worked out such methods by watching the herring-gulls which often smash shellfish open this way.

When Iain next came by I told him about it. He felt the hoodie was as cunning as a fox. 'D'e know it knows the difference between a rifle and a fishing rod?' he asked. 'If ye want to shoot a hoodie, disguise your gun like a rod and ye'll fool it. Once I saw one sneaking along like a duck through a corrie to avoid me, keeping a large rock between us, so as not to be shot at. Then it flew from a peak further away. Some of them *know* the range of a shotgun.'

Now April was doling out two more weeks of winter weather and hailstorms still beset my boat trips. Once the engine handle came right off and I had to hold the throttle with a pair of pliers, my hands frozen. The upper cylinder was showing wear, the plug oiling up frequently so that I had to limp back on one cylinder while flailing away at the oars. I had to shelter at the south-east bay until the tide was high enough to go through the gap between the mainland rocks and the seals' islet to reach my own shore rather than face the rough, open sea with a faulty plug.

I used the two-hour delay to gather loose cockles on that beach. As I climbed from the boat I startled what seemed to be a huge hind which shot to the skyline where it stopped and looked back. It would have been easy to shoot. As I trained my field glass on it and saw its massive body more clearly I thought it was a stag that had dropped its antlers. But I could see no sign of the white, blood-streaked pedicles on its head which would have been normal. It was certainly a 'hummel,' a permanently antlerless beast.

The growth of the large red deer antlers, which are shed early each April and grown again in a year, is still not completely understood. While it is believed they may be a vestige from the days when deer had to fight off bear, lynx and wolf, the last of these predators, wolves, were exterminated in the 1740s. Meanwhile antler growth, when the food supply is good, has not significantly diminished. Certainly stags use them for occasional jousting – more pushing and shoving matches really – when defending their harems of hinds from other stags in the autumn rutting season, but very often they fight with their forefeet, rearing up into the air to deliver powerful downward jabs or outward kicks. In this way a hummel that does not have to suffer the annual drain of calcium and energy that goes into growing antlers, can often increase to a huge size and finds no trouble in becoming a master stag.

One day, after writing for some hours, I found the hazy sun too tempting and went to watch some red deer on the far ridges above Shoe Bay. The sun was almost behind me and the wind from the west – from them to me – ideal conditions. I decided to try my first real deer stalk on Shona. Squeezing like a lizard between the heather and crawling crabwise over flat rocks, my hands, knees and elbows were soon wet as I went through gullies. I tried to keep heather or grass before my face and peered slowly after each sequence of movement to see if I'd been spotted before moving on. I came to within forty feet of them before reaching an impossible corrie, a dished hollow between myself and them. They were certainly easier to stalk than the British Columbian mountain caribou. Their eyesight seemed poor, except for detecting actual movement, as if they could not make my head out from old blackened heather, which no doubt it resembled. But they were now suspicious. A ewe had barked a warning from high above and one large hind, redder than the grey-brown coats of the others, walked round to my right, slightly more windward. There were five hinds and two young staggies with tiny six-inch spikes. As I watched the moving hind through my field glass I thought what knock-kneed grace she had, such slender legs and delicate unfurlings of the foot as she moved. I saw her faintly reptilian head, buff muzzle, black nose and huge hazel eyes like marbles. How I wished then I had my camera back.

Then, for the first time, the idea struck me – why not also photograph wildlife in the future, as well as trying to write about it? Surely I should be recording all I saw around me, and even a complete novice like myself could have taken good pictures on this day.

Finally I raised myself up on two fists and stared. They did not run but just stared back, then started moving after the hind.

Even in late winter my first sighting showed an idyllic scene.

No South Sea island could boast anything more beautiful than Shoe Bay. It was an otters' playground.

Returning from a trip to replenish supplies, with timber, fuel and food.

After boating everything to the island, it did not take long to replace the croft-house walls.

Harry Heron – whose life I saved by putting him back in his nest after he had been thrown out by a storm

In winter a plastic tent round my desk kept in all the warmth from the oil lamps and a tiny paraffin stove.

When they came to know me, I could feed the seals with surplus fish. I freed the young seal with its front flipper caught in a crevice after which Sammy, more tame than the rest, would often follow my boat.

Buzz, the young sparrowhawk, had been found tumbled from his nest.

Although his pinions were not yet fully grown, Buzz could easily catch the food lures I pulled over the grass, clinging on with surprising strength.

On fine days Buzz loved a daily bathe in the burn, shuffling water over himself with his wings.

While all the females are incubating eggs, the resplendent eider duck drakes form seaside social gatherings.

Herring gull chicks in three stages of hatching: one is out, another half out, while the third still chips away at the shell.

Strong lambs would leap across my burn.

My first view of young Sebastian.

A successful stalk!

In the winter cold hinds came lower to graze where the relative warmth of the sea helped to melt the snow.

Despite a broken wing and injured leg, Gilbert the herring gull clung tenaciously to life and after more than two months of nursing recovered his health and went free.

Hoping to be free.

Free!

A large tom wildcat, showing its usual reaction at the close presence of a human.

It was idyllic, just sitting there fishing in the intense blue-gold beauty of that wonderful seascape.

Clearly, if you don't look like a man they move around to try and get scent, to find out what you are. They began to walk *towards* me, intently curious. The curiosity of deer, if stalked correctly, has often been their undoing. I stood upright very slowly. Still they stayed. Slowly I turned my back and retreated, and they followed after me for almost a hundred yards, so that I began to feel they were actually stalking *me*. To date it was my most unusual experience with wild red deer. Darn that burglar – for I'd heard nothing from the police about my lost possessions. That night I wrote to a photographic firm I knew of in London.

Spring was really under way, the male eiders now cutting off their siren mating calls before the last high note, zooming around and chasing the ducks on the sea's surface in flapping displays. Near Deer Island I saw a pair of ornate shelducks swimming. With their red beaks, green-black heads, bright chestnut chestbands and white necks, backs and bellies, they looked like a couple of carnival tugs steaming along. They often marked time in the shallows, treading up the sand into billows and upending to grab whatever morsels of food they had stirred up. Britain's biggest ducks, they reminded me of geese and, when they flew, the beats of their black-and-white wings were also regular and slow. Now the little black guillemots, or 'tysties', bobbing like burnt corks on the rippling waters, were courting on the loch, swimming round their mates, giving squeaky cries and showing their bright red throats. If my boat went too near they flew up, hitting the water with feet and wingtips with a noise like heavy pattering rain, until they became airborne and whirred away like mechanical toys.

On 19 April I saw the first worker bees humming from flower to flower, and a queen wasp, looking for a nest site, landed briefly on my hand. A tortoiseshell butterfly, fresh from winter hibernation, flicked from dust patch to dust patch, sunning its wings, and just above me the dark underside of a comma butterfly went past in typical spiral flight, just as they had done over the hazy Sussex ponds of my boyhood.

Iain came halloing over the hills next day in bright sunshine and we found a dead lamb lying behind the old ruined croft.

'It was stillborn,' he said, looking with expert eye at the whitish fluid and blood of the afterbirth. No fox had had it for there was no tell-tale scarlet slash on the throat. 'We'll have to bury it, not through sentiment, but if ravens, hoodies, buzzards or black-backed gulls see it they'll be after pecking out its eyes. Once they get the taste they may go after sick or new-born lambs. We don't want any more killers here. I've found four lambs killed by foxes already.'

I prevailed upon him to let me lay the lamb out on a grassy patch three hundred yards south-west of the croft, so I could see what bird came down first, and promised to bury it later. Although I kept watch from my desk all afternoon, and one pregnant ewe sniffed it without concern, the only birds that showed interest were a pair of greater black-backed gulls. One looked down, alighted near the corpse, circled twice on its long black white-edged wings, stayed still a few seconds, then flew away again. Perhaps it was too near the croft. I buried the lamb in the gathering darkness.

Two days later I smelled burning. Outside the air was filled with smoke. I rushed out and in a dell a quarter mile away found heather and old bracken ablaze, smoke puthering upwards in great brown clouds. I hurried over to inspect. The fire was creeping in the west wind towards me and the croft! Yet there seemed to be no smoke anywhere else on the island. What on earth was going on?

I ran to the croft for a bucket, then back again and tried vainly to douse the flames with water from a nearby burn but I had to give up as the fire leaped from one heather bush to another. I stood there in the burn, determined to fight it again should it cross the small gorge but the fire finally died on an open green patch a hundred yards away. I threw more bucketfuls along the bank on my side in case it started up again, and went home.

Later, as I was sunbathing, Iain appeared and said *he* had started the fire! 'We always burn the heather when it gets too long every few years,' he explained. 'Young heather is good for sheep, especially in winter, but once it's over a foot long they won't go near it. We burn it down and grass grows before the new heather. This is also good for the sheep. Some of the heather wasn't burned even in my uncle's lifetime.'

I was amazed. What about bird nesting grounds, all the flowers that are needed by valuable insects like bees that pollinate other plants, the eggs and caterpillars of some fine moths, and all the mice, frogs, lizards, insects, beetles and other small wildlife so important in the web of natural life, and on which many larger species depend? And what about all the young regenerating trees? What if the fires went out of control? Iain seemed surprised at my questions. Heather burning was traditional in sheep country and in the management of grouse moors in the Highlands. As for control, he always tried to burn only where there was a natural end, like an open area, gorge or rock face.

In later years I was to find that, unlike Iain, some hill burners made little attempt to control the heather fires and often, by accident, whole areas of the Highland hills would go up in flames in April, burning out many square miles at a time.

It seemed an oddly short-sighted policy. Such burning does not, for instance, affect the underground roots of bracken which quickly spreads into new areas, restricting new grass, flowers and slow-growing heather. W. H. Auden's phrase when describing the Scottish mountains came now to my mind – 'Hills like slaughtered elephants'. Trees and plants are, along with the phytoplankton of the seas, the lungs of our planet, and the forests and bushes keep the watershed up. When such burning goes out of control all young tree and bush seedlings are destroyed, and when too much heather goes also, only sparse grasses and sedges, in brackenless areas, remain to take over. On steep hills exposed to heavy rains and gales, erosion occurs rapidly and then even the earth for growing grass is blown and washed away. Some long heather is useful shelter for sheep and birds like grouse. Heather burning can surely only make sense if done in very small areas at a time and heavily controlled, which means costly manpower, experience and ecological good sense.

Two mornings after the heather burning, I saw an oyster-catcher bathing again in four inches of water in the mouth of the burn. And watching the oystercatcher was Harry Heron, who was standing right next to a herring-gull on a large flat rock. The two were only about a yard apart. How odd, I thought. Had they formed some kind of friendship? The next thing I knew the oystercatcher was actually swimming, something else I had never seen before. It was then I recalled my vow to get a new camera. But for such small species as birds, wary in the wild, I'd surely need some telescopic device. Well, I would get a telephoto lens, whatever that was.

Towards the end of April more wildlife was stirring and only the oak and ash trees were not yet sending out new leaf sprouts. Long-horn beetles were out in the sun looking for mates, and the first bumbling black dor beetles, which help bury carrion – in which they lay their eggs – were burring past in the air. On a walk along the path, treading down the new snake-like bracken shoots which, like all forms of life, are easier to control in infancy, I saw two mating coach-horse beetles. They were joined at the rear and the slightly larger female was towing the male about. He seemed to be trying to help her movement, or possibly to ease his pain, by desperately thrusting backwards with his legs in the direction in which he was being hauled. That seemed to be taking women's lib a bit too far!

Near the beach I saw the first wheatear of the season. Handsome with its blue back, black wings and face mask, chest-nut breast and white rump, it seemed to be flying weakly, trailing its wings as if to decoy me away from a nest. It was far too early for it to have bred and was probably exhausted after its spring migration flight from wintering in North Africa. It fluttered feebly between two rocks a hundred and fifty yards away and stayed as still as a statue, watching me with what seemed excellent eyesight. The wheatear has become rare in southern Britain (where once it was common) due to the ploughing of downland and the loss of many heathland habitats, and I hoped it would be all right.

But winter had not given up yet. Three days of hail and darkness came, rendering the landscape barren of activity. Then, as the sun beamed down on 28 April, I was woken by the loud calls of the cuckoo – so close I thought it must be on the roof. When I peeped through the window, it was actually in the ash trees a few yards away. I had never been so near this wary bird before and I was astonished by its beauty. With the sunlight burnishing its slaty-blue back, long white-spotted tail, grey-and-white barred chest, slightly curved beak and the glint in its bright yellow eyes, it reminded me of a sparrowhawk.

Again I cursed having no camera. The photographic firm in London had said they would give me a discount. Well overdue for a work trip south I drove down to London and arrived back with a new 35 mm Pentax camera, standard lens and a cheap 300 mm telephoto lens. For the first time I made it back in one long drive, using no ferries at night. When I reached the little pier in the loaded old Land Rover there were 573 new miles on the clock. I felt more tired than a rally driver.

As always the sense of loneliness was greater after returning from a trip to civilization. I used the wet days to work on the Canada book, re-creating the wild bear treks, and walked the Shona hills in finer weather. The Highland landscape seemed more than food enough for my heart. That fine writer T. H. White had young King Arthur's tutor Merlin say 'The cure for being sad is to learn something new'.

Well, after all, I was now not only watching nature, I was also a wildlife photographer!

9

There's More to Wildlife Photography

During the following months I was to discover how little I knew about wildlife photography. By some obscure magic, glimpses of wild creatures ceased almost completely. The buzzard and kestrel, like a tiny swallow beside the bigger bird, still flew over but seemingly higher up, and it was hard to focus the telephoto lens in mere seconds. They looked fine in the viewer but when the pictures were developed the birds resembled mosquitos squashed on a window pane, their wing tips or heads cut off by the edges of the photos. It took a long time to learn that exposure had to be related precisely to the object I was photographing and that sometimes it was necessary to work against the meter reading, closing the aperture a stop or two or increasing shutter speed in sky or sunny day shots to avoid overexposure.

When I went to the heronry on Deer Island I found Harry's low nest had been partly blown out of the tree and had to be content with shots of him and other herons flying over. All the other nests were so high I could neither climb up near them nor build the essential tree hide without help. I found a shelduck's nest with ten creamy eggs beneath an old holly stump but though I hid under camouflage netting in a bush for several hours, the parents did not come near. I gave up in case the eggs grew cold.

Although Charlie Crow and his mate were still coming to the bay for marine prey they did not use their old birch tree nest. Maybe they had an inkling of Iain MacLellan's intentions. After a long search, I found their new nest in a tall ash tree fork in a wood to the south-east. It was made of dead twigs as if to make it look ancient and deserted. Now, in the breeding season, Charlie no longer trusted me either. While I was still a good way off his mate, incubating the eggs, would slip noiselessly away, dropping down low and swerving through the trees. But one morning I saw him fly between two large rock folds and not come out again.

I stuck herbage and dried seaweed in the netting of my bush hat and feeling like I had during army sniper training years before I stalked him, crawling through the wet weed between the rocks. Hooded crows, in their distinctive black-and-grey plumage, appear to have poor camouflage for their nefarious hunting methods but I could not see him. Finally I did – he was working through the tide wrack for sand hoppers and tiny crabs and his grey waistcoat matched the rocks while his black back blended perfectly with the dark brown of the seaweeds. I managed one photo of him but when he heard the click of the camera he flew away with indignant '*Kaahs*'.

Two days later, with the Canada book well under way, I took the afternoon off for a long photo trek. More humiliation was ahead. First I walked to Shona House for my mail. Seeing no wildlife at all, I then headed over the highest hills to the north shore, to make a long sweep of the island back to the cottage. It was a muggy, clammy day and what little wind there was kept switching – from east to north-east, then north to north-west. From just below a high peak (never show yourself on the skyline) I saw a herd of red deer hinds a good mile away to the west. I stalked them, keeping close to rock faces through gullies and bogs, coming nearer to their ruddy-brown grazing forms. Setting the telephoto lens at correct aperture and speed, I slid round the final rock and took my first deer photo – of the buff, heart-shaped patches of their rumps as they ran away! I learned again, more strongly in these open hills than I had in the wooded mountains of Canada, that on moist days scent carries further and that stalking is often hopeless in changing winds.

From Sgurr an Teintein (High Peak Of The Deer) I traversed the yellow hill of Cruach Buidhe and the four hills to Aonach and

Shoe Bay but found only three winter-dead sheep, the carcases ripped apart by foxes and the beaks of predatory birds, with white wool scattered over many yards. I photographed a bloody skull near a rocky den, which may have been used by a fox as a sheltered eating place for it was too shallow for a breeding earth. High on grass tufts and small rocks foxes had left their usual dark-grey hair-filled 'visiting cards', as well as urine to give information on sex, mood and boundary to others of their kind. Only one seemed fresh, filled with blue-black wing cases of beetles the fox had probably taken from beneath a cleaned-out carcass. I approached Shoe Bay stealthily, but not one otter did I see.

Back home, perspiring after the five-mile trek, I bathed in buckets of water and learned that in the burn itself I had an ideal washing machine. If I put my clothes into the rushing waters two hundred yards above the croft, they were beaten clean by being tumbled over rocks and little waterfalls, and I could pick them up in my main pool ready for soaping with warm water! I reckoned natural organic dirt and sweat would do the burn no harm but I drew the line at putting detergent into its waters. I just rubbed the clothes in buckets and threw the pollutive remains over the virulent bracken roots.

So my first real wildlife photo trek in Scotland had proved fruitless, but I found a compensation in my mail. The American colour magazine that had printed the story of days on the salmon boats in British Columbia now wanted a 'philosophical' article about life alone in the wilds, and I could 'let it run'. At last a fine publication was interested and during the next week I felt inspired, truly alive. At last I was starting to *contribute* a little and writing what I wanted to write.

Needing exercise after long hours at my desk, I dug more stones out of my garden soil and checked the lambs in the area for Iain. One morning I found a hefty youngster limping badly on its right forefoot near the south shore and tried to catch it to see what was wrong. It looked ready to drop yet it could run like a deer on three legs. I chased it for half a mile and was about to throw myself on it when it reached an area of scrub trees. Too late. Then, after sneaking up behind some bushes, I made a mad dash, threw myself down, clutched its thick wool with one hand and struck the side of my face on a rock. The lamb had a small stone wedged in its cloven hoof. The foot was sore but not bleeding. I prised the stone out while the lamb panted feverishly in my ear and then let it go again. When I got home I found my own wound was worse than the lamb's, with blood trickling down my neck. Now I *knew* why Highland shepherds used fleet-footed collies!

A couple of days later I heard high-pitched bleating behind the croft. I went out to find a mothering ewe had leaped the burn and was walking on, her little late-born lamb still stranded on the far side, looking helplessly at the yawning chasm separating it from its mother and giving pitiful cries. As I watched it tried to emulate the ewe, took a jump, and landed straight in the burn, *splosh*. Now it was up to its neck in water and couldn't climb the high slippery rocks. It set up a frightful bleating but its stupid mother, not watching or apparently hearing it, just carried on grazing a few yards away.

I rushed over and lifted it out, wringing the cold water from its snowy fleece as best I could. It cuddled close in my arms warming up from my body heat and snuggling its head up close to mine, totally trusting. I had to resist the temptation to take it back to the cottage as a pet. But the moment I let it go again it ran bleating back to its mother. Next morning the little lamb spied me walking up the slope from the beach and suddenly ran towards me with joyful little '*meheh-ehs*'. I picked it up, noting it was a little tup, and as I stroked its head for a few seconds it closed its eyes blissfully. Then I had a sudden guilty feeling. If Iain spotted me he might think I was alienating his sheep from their natural life. When I set the lamb down, it ran straight to its mother again, making high, skittish, head-wagging leaps on the way.

That afternoon Iain came across the hills carrying a dead lamb and showed me the tell-tale crimson on its neck. He was angry that the big fox was still around and said the hounds and some terriers would be back in two days.

'We were unlucky last time,' he said. 'I saw the vixen go into the den. In went the terriers, got two cubs, but didn't bring them out, and we had to scrape them out with sticks. And they wouldn't go in again, afraid of the vixen. When terriers have been hurt in fights with many foxes or half-killed by a badger, they're not as keen to work as when they're young and fearless.'

I asked him if he thought wanton killing by foxes might be linked to revenge. When a fox had its mate or cubs killed, would it try to get its own back, knowing lambs to be the products of its human persecutors? Iain agreed there might be some truth in the idea but said that foxes often killed like that before hounds or den terriers had been at them.

I told him about the little lamb in the burn and he laughed. 'I've known it happen too. I picked one little fellow out of a burn with the afterbirth still on it. I put it on to the bank, but as I was getting out and calling to my dog, blow me if it didn't leap straight back in again. I put it back to its mother but the right smell had gone because of the water and she butted it away so hard she sent it flying from here to there. She was a poor mother anyhow, and in the end we had to take the lamb away from her and rear it on the bottle.'

Once he found a lamb that had fallen down a crevice. 'It was a tup and a good job it was for it had a wide head and its little horns had wedged into a gap between the rocks. I had to get below and lift it out. The little thing was so far gone, at first it just staggered about and rolled head over heels.'

As the soft, soothing notes of cuckoos were now sounding all round the croft, I told Iain of the one that landed near my window. It had looked like a sparrowhawk, and he told me an odd tale.

'A sparrowhawk come through our window when we were having lunch a few days ago,' he said. 'It must have been going at 60 miles an hour for it made a circular hole. We heard a crash, then it landed dead on Morag's plate, its skull and ribs shattered. One piece of glass shot right across the dining room and kitchen and landed in the sink!' He wondered if it had been chasing a small bird which had dodged away at the last moment. I said I doubted that because, with its long tail, the hawk can dodge as fast as any bird. I knew Iain had a canary and wondered if the hawk had seen it through the glass – or if it had seen another bird outside the kitchen window. 'Maybe at speed, not seeing the glass, it thought it could give a quick flip of its wings inside the darker room, then close them when going through the far window to strike the bird.'

But Iain said the canary was in the corner behind the wall and the windows were not arranged so the hawk *could* have seen straight through! In the end we decided the hawk must have seen the reflection of the open country *behind* it in the glass and thought the way was clear. At any rate, Iain said, the poor sparrowhawk had been buried by his elder son James with full military honours.

When Iain left I set the dead lamb on a grassy patch by the shore and kept watch from my desk for two days. A hoodie flew over, circled once then kept going. A kestrel hovered over it briefly, decided it wasn't food and flew on. Only a greater black-backed gull showed any real interest, standing for an hour on a nearby rock, but by dusk nothing had actually touched it. Not wanting to encourage foxes, I took down my shovel and buried it, as I had the first.

Next morning I worked over my garden soil, sieving out the last stones and piling them with the other rocks I had heaved out into twin windbreaks at each side. Rain began to pour down, and after digging a drainage ditch too, I went indoors. Suddenly, at dusk, the air outside was filled with whirring wings. A flock of herring-gulls had come in from the shore and, attracted by the snails that had emerged from hiding and the worms rising to the surface after the rains, were walking among the bright-green bracken stems like stately white ships.

I was even more surprised when a pair of oystercatchers joined them. These gaudy seashore birds probed their three-inch-long beaks into the earth like thrushes and pulled out large worms. For the first time I noticed their eyes were near the top of their heads, giving them virtually all-round vision. Oystercatchers, it seemed, were the woodcocks of the sea. The gulls did not return but on rainy days for almost a month the oystercatcher pair came whinnering around the croft, always at dusk and pleeping shrill penetrating cries if I went outside. Once I had been astonished by an oystercatcher flying alongside the Land Rover, keeping up at 45 mph as I was driving over the high Beattock Summit. It was almost thirty miles from the nearest sea at the mouth of the River Annan, yet the bird had surely flown up that valley.

It seemed I was witnessing evolution at work. Gulls (such as the common, black-headed and herring) have been coming inland for years, attracted first by ploughing on farms near the coast, then to rubbish tips near towns, and in London I'd had flocks of herring-gulls come round my bed-sitter building when some of us tenants put food out regularly for them. Here it seemed, with the coastal seas being heavily fished, more shellfish gathered by humans and seaweed cut for processing, that oystercatchers were also catching on to the trend and turning to the land for new food supplies. It was the opposite of the hooded crows' graduation to the sea shore to exploit, for them, a new ecological niche.

I knew that many of Britain's twenty to thirty thousand oystercatcher pairs wintered in the north Irish Sea and along the Welsh coast (with birds from Iceland and the Faeroes) and fed in the cockle- and mussel-producing areas like Morecambe Bay, the Wash and Burry Inlet, where a campaign to shoot 11,000 over the two winters was launched by fishing interests in the early 1970s. They were known to forage in fields as well as mud flats in the winter but this was the first time I'd known them feed inland in early summer. As I watched them it became clear they were not only eating worms but also performing a useful service by taking many leatherjackets (larvae of the cranefly and a pest to plants).

Later I was to find out that oystercatchers going inland to feed and breed in the Faeroes frequently entangled their legs in sheep wool until they could no longer walk. A vulnerability, perhaps, from not having learned over many years' experience to step high and avoid the wool in such new foraging grounds.

By 10 May the Arctic terns, 'sea swallows', were back from their amazing 10,000-mile migration flight from the Antarctic, flipping over the sea with slow but strong down strokes that jerked their slim, little grey bodies a few inches higher with every beat. They often poised in the air, tails expanded like hovering kestrels of the sea, then with beaks pointed downwards, plunged in with a slight twist and half-closed wings to catch tiny surface-swimming fish. Then they bobbed back up again, took off, shook their wings with little vibrating quivers to spray off the excess water, and carried determinedly onwards.

Green-veined white butterflies now filled the air with their flittering forms. There seemed to be fewer herons about this spring, but Harry still came wafting into the bay at dawn to stand deathly still like a pale blue-grey ghost, silvery incandescent in the rocky shallows.

While I was digging the garden on 18 May, glad that the first midges seemed unable to bite so often or as strongly as the late summer ones, I noticed the tardy ashes were at last catching up with the trees around them, their thickening brown buds sending out their first leaf shoots. It took two more days to get the garden ready for planting. I spread two sacks of sheep droppings, which I had dried and pounded into dust, over the surface and then threw on many more sackfuls of the fertile soil from the new molehills scattered around the croft. After a final raking, I planted carrot, cabbage, lettuce and radish. Taking a tip from Iain, I put flat stones between the rows. These catch the sun, hold heat long after nightfall and thus act as radiators.

Next day I built a preliminary fence round the garden with old sidings from the ruin above, and within nine days the first radish and lettuce shoots were coming through.

Returning from the pier a few days later, after a quick trip in the rubber dinghy to post off the movie book proofs, I heard a loud plapping noise, like an elephant breaking wind, and down went the port side of the boat, dumping me in four feet of water. I had passed too close over a mussel bank and the hidden edge of one protruding shell had ripped a 17-inch gash in the boat's flank.

'The mussels' revenge,' I told myself as I waded ashore with torn boat and wet supplies in the south bay. But as I laboured with the engine, oars, jerrican and fuel over the ridges to my own bay I remembered: I'd never taken any mussels from that particular bank! At dusk I rectified things. I went down and carefully removed all the offending prominent mussels and made the best *moules marinière* I'd ever had – as compensation. The £60 boat was ruined. When I saw that the rubber was only paper thin, I re-learned an old wilderness lesson: never buy cheap. You can lose your life trying to save a few pounds.

An attempt to repair the boat with lorry tyre-patches failed when the sun peeled them off again. Finally I bought some liquid polyurethane, shook the two requisite mixtures together in a large plastic bottle and squirted the rapidly expanding and solidifying contents into the hole between the patches. If the other two air compartments also became punctured I would end up with a rubber boat more solid than the main fibreglass one, and totally unsinkable too. Meanwhile I had a boat that was merely one-third solid but it was almost as buoyant and went as well as before. Now if the other two chambers went flat in a sea disaster at least I had something to cling to while getting ashore.

One hot day in May, the first totally cloudless day for a month, I went to take my first photos of the seals which basked in the sun on the rocks just off the mainland peninsula across the mouth of the loch. I rowed over and hauled the boat on to a beach of fine shell sand, so blazing white in the sun that it hurt my eyes.

The seals had hunched into the sea at my approach, so while they settled down again, I trekked round the peninsula and saw several herons rise one after another from the flat bowls of peat bogs. Wondering what they were finding, I walked over the yielding blanket and examined small pools. Tadpoles! The little black commas wriggled in every patch of water but not a single frog did I see. With so many gulls and herons about perhaps the adult frogs had bred a more devious instinct than usual. After an hour, seeing little more than a huge pair of greater black-backed gulls flying overhead with deep '*qwuck qwucks*' at the rare intrusion of a human, I went back and stalked through the heather to the small cliff to photograph the seals.

They lay on the rocks, with their wedge-shaped flippers sticking out or closely tucked into their sides, like great shiny moles with snub, up-turned noses. Two of them were very fat, as if about to give birth. Their colours varied greatly, the larger ones being blackish or mottled brown, while smaller ones I judged to be eleven months to two years old ranged from light brown to yellowish to grey-white, and there was one marbled with dark grey and cream that looked almost blue. All were common seals (a misnomer because in fact the larger Atlantic grey seals are more common) and their attitudes and postures were hilarious. There was a large brown one that lay on its side like a huge sausage flipped from a pan, a black one on its back, eyes tightly closed, velvet brows furrowed like a bloodhound's as a slight bump in the rock pushed its upper neck skin forward. The small, marbled grey seal held one flipper straight up into the air as if saying 'Hi, I'm over here'. Two had their heads and rear flippers curved upwards as if doing back exercises.

I took a few shots, disappointed that I couldn't go closer because of the open water between the land and their islet. I could hear odd snorts, snuffs and digestive burps that sounded like growls. A noisy lot! After forty minutes I was surprised when two common porpoises came in from the open sea and started their typical undulating surface swimming, comma-ing in and half out of the water between and round the islets, and then to see some of the larger seals hump themselves into the water and start to 'play' with them. Certainly there were no attacks of any kind and it seemed the porpoises were welcome friends. Two of the larger seals appeared to imitate the porpoises' style of swimming, like a human butterfly stroke without arms, their two rear flippers flailing away on each shallow downward dive, and I managed some photos of this.

I had read that bull seals will perform this 'salmon leaping' swim in the mating season, but that wouldn't be until late July or early August. Common seal pups are born in this area from early June and the females are ready to mate about six weeks later. The pups are born on the haul-out islets or mud- or sandbanks, can swim at birth, and suck their mother's rich milk for a month or longer, both above and below water.

After two hours I slid backwards through the heather and went to the boat. On the shallow beach the low tide was now so far out from the mouth of the little inlet that the sea had shrunk right out of sight! The boat and engine were almost immovable on the high friction sand with no wet planks to haul it over. I had not brought any lunch with me. Oh well, no time need ever be wasted in the wilderness life. I collected fifty-one cockles for pickling, still leaving plenty behind. I had been worried that my taking cockles would depopulate the local species but it soon became clear that only a small proportion came up on the tides and most were left in deeper waters.

I collected some mussels too, realizing that day that the best were not necessarily the biggest, or those closest to shore. Best were the mid-sized two-inchers that were *always* covered by water, even on the low spring tides. The shore mussels, exposed for up to half their lives by the tides, grew more slowly. They were often heavily encrusted with barnacles, and because of the action of the waves on the sea bed around them, had to cover specks of sand and grit that got between their shells with mineral deposits. If you didn't chew such mussels very carefully they broke your teeth. I kept dozens of these 'pearls' in a matchbox and once, when I lost a filling this way, I found one that fitted the hole exactly and used it to plug the gap, with cotton wool dipped in toothpaste, until I could reach a dentist some months later. From the day I visited the seals, I dived with a face mask at low tides for the better mussels, a linen shopping bag tied around my waist. I had no fears my small needs would decimate populations. Long beds of thousands of tiny mussels less than a quarter-inch long, with their furry open tips filtering the plant plankton from the water, looked like soft, blue-black carpets. Man can live in the wilds without causing harm – it is always a question of timing and degree. And it requires a humble mind.

I was surprised once to read a news report that heavy conglomerations of mussels were encumbering girders on some of the new North Sea oil rigs. To get rid of them the crews cut channels through intervening beds of sea anemones, so that starfish (which prey on mussels) could get up to them. Apparently starfish either can't climb over the slippery anemones, or won't because of their stinging tentacles. But why didn't the crews harvest mussels for their *own* food?

Early June provided some freak weather. Working on the croft's west wall and window, I boated to the cabin site for some timbers and was hit by a sudden hailstorm that seemed to sweep out of nowhere over the hills from the north-west. I was shivering in my soaked short-sleeve shirt, glad of the warming exercise of carrying the lumber up to the croft, when I saw the oystercatcher pair again. It seemed that when heavy wind made waves pound on shore they found it easier to hunt for worms inland. I realised too that, if they fed their young on worms and caterpillars, the chicks might thus acquire a taste for them when adult too, and in this way the evolutionary trend would be compounded.

On 7 June I was standing ankle deep in the sea on the west side of my bay, counting the first beautiful yellow flags that were out in the marshy area by the burn's mouth, when I saw three oystercatchers, a curlew and one hoodie working the tide wrack for food on the far side. Suddenly I was surprised by a large plaice that shot from just below my feet and went ribboning out to the bluer, deeper waters.

I realised then what a great part the tides themselves played in sustaining many forms of marine life. On every rising tide such fish as plaice, dabs, flounders and rocklings came gliding up over the crabs trundling along the sea bed to scavenge whatever food – dead insects, deer flies, bluebottles, gnats, spiders, grubs and beetles – had fallen off trees or been left on the beach, or had drowned in small pools while the tide was out. Just as foxes, ravens, crows, gulls and other birds came to feed on dead fish, crabs, sand worms, cockles, winkles, whelks and other shellfish and molluscs broken loose by the pounding actions of the waves when the high tides retreated again. It was a two-way swap, twice a day, that played a great part in the whole balance of shore life. Even the size and strength of the shore birds' beaks fitted each into subtly different feeding notches. Curlews, for instance, with their nebs up to seven inches long, can probe for tube, lug and white worms, or razor shells and bivalves that burrow too deeply for birds like oystercatchers or gulls to reach. But the thick powerful bills of the oystercatchers can pluck out the shallower burrowing cockles and stab at the adductor muscles of the filtering mussels, two of their favourite foods. They, along with crows and gulls, can also knock limpets from rocks with a sudden sneaky blow. Most shore birds – including eider and shelducks, guillemots, and turnstones too – can also feed on other molluscs, nereid worms, shrimps, young flatfish, sand eels and rocklings. A favourite food too was the sand-hoppers which often leaped in small clouds, thousands strong, when feeding from the jetsam on the tide wrack at dusk.

Next morning I had a brainwave. I spent the day making a plaice trap out of a fish box, covering the top with half-inch wire netting and fitting a hinged fringe of thin wires at one end. Holding it down with rocks on the sand, where it would be covered with ten feet of water at high tide, I baited it with bacon rind, worms and mussels. The idea was that flatfish would scent the food and sneak in over the slitted bottom through the inward-hinged wires – and not be able to get out again. A good idea, but it caught not a single dab! Tiny crabs which nipped in and out of the netting took all the bait. The hinged wires were just too roughly made for a flatfish to push against them. For such a trap to work it would need to be machine-made with great precision – an ironic admission. However, I did manage to catch a few prawns with a funnel-shaped nylon-mesh trap baited with fish scraps and held on the bottom with a rock, with a float attached so I could haul it up.

One afternoon, after being roused from a sunbathing nap by the shrill '*pleep pleeps*' of the oystercatchers, which I presumed were making them for devilment as they had no nest near the croft, I stood in the deep pool of the burn to photograph the birds. Suddenly I saw what looked like two branches nodding up and down behind heavy bracken. A large stag with widespread antlers, which in early June were not fully grown, was casually grazing sixty yards north of the croft. The antlers were in full velvet and it seemed extraordinary they should have grown to that length since I'd last seen the stag with mere twin knobs back early in April. I raised the camera slowly and took two photos as it stared back at me arrogantly, seeming not at all nervous. It was now in its fourth or fifth year and looked a superb specimen.

I allowed him to walk out of sight before moving myself. I tried to track him over the ridge to the north-east, in the hope of stalking for more pictures. But with only brief rain in the past four days his slots were shallow, and I lost them on beds of dried old bracken, the grass and rock faces and saw no more sign of him after going down to the south shore.

I need not have bothered for he was back the next two mornings, exasperatingly around dawn when the light was too poor for photos. By the time the daylight was strong he was away in the hills. I took to regarding him as a wandering itinerant neighbour, the kind who is never there when you want him but who helps himself to whatever he needs of yours. When once I saw him staring at me with a wedge of herbage sticking from the side of his mouth like a pipe, I nicknamed him Sebastian. Usually he walked slowly away without showing the panic common to his kind. Every time he stopped he would turn and gaze at me as if saying, 'Don't look at me like that. This is my place as much as yours!'

One day, as I emerged to get a bucket of water and startled Sebastian, he took a spectacular flying leap across the burn and galloped away. He had nibbled all the tender new green shoots from the bramble bushes that supplied my autumn blackberries. Now I knew also where my dock leaf 'cabbage' was going. When I saw Sebastian's large backside, with its heart-shaped light buff alarm patch and ludicrously short tail, a mere three feet from the window, I felt he was taking things too far. His other end was busily cropping away at the cabbage and lettuce plants in my hard-won garden patch.

That morning I boated up the loch for the old sheep netting below the big house that Iain had promised me, and for a large section of blue, nylon, fish netting which I had seen drifted up on shore opposite Deer Island. With these I made a stag-proof roof for the garden.

10

Encounters with Seals

Knowing that good luck comes in the plural as often as bad, I rowed one morning to an islet on the loch's mouth where many sea birds were flying. To keep my camera and lenses from being splashed by spray I wrapped them in plastic bags. I had walked only five yards on the bright-green turf amid the colourful sea pinks and grey-and-white lichened rocks when a brown eider duck shot upwards from the vegetation and flew away. Her grass nest, lined thickly with the soft brown down she had plucked from her breast and belly (its heat insulating properties are employed in eiderdown quilts), contained only three light-green eggs, one of which was hatching, the egg tooth of the duckling's beak showing through a small hole. Perhaps she had lost other eggs to crows or foxes, or even otters, for there were fish bones on the islet's rocks.

Two herring-gulls still had three eggs in their nests but four other pairs had hatched their young out a week or so earlier. The downy-white and grey-speckled chicks squeezed tightly into crevices when they heard their hovering parents making loud '*qwuck qwucks*', a signal for them to hide. When I picked one up, it responded with a look of fateful innocence. I also discovered an oystercatcher's nest, a mere scrape containing pebbles the same size as the three eggs, as if to help camouflage them better. These were smaller and more darkly blotched than those of the gulls. I had an idea I might find one, for the parent birds were flitting round the islet in total silence.

They call loudly when one is several yards from the nest, as if to warn chicks to hide, but go quiet again when one is very near it. I took my photos quickly, making sure not to disturb any nest or vegetation, then quietly left the islet.

With the boat engine now breaking down frequently, I rowed to the store at the head of the loch to shop and post mail. It was a long ten-mile haul there and back, and while my hands were not blistered, my backside was sore from rocking to and fro on the hard seat. As any distance oarsman knows the rear takes more punishment than the hands once they've hardened.

A few days later I went back to the islet to photograph the sitting eider. Remembering the exact spot, I stole quietly near and took a picture of her head and neck in the deep bowl amid the long grasses. She did not move and I backed away to leave her alone. As I stepped sideways another eider on a nest I had not found earlier shot upwards. I looked down – she had covered her eggs with a squirt of thick, greeny-brown liquid that smelled faintly like a fried bacon, liver and mushroom breakfast! This was not her excreta, I found out later, but a special fluid eider ducks emit to camouflage their eggs in an emergency, its scent probably being abhorrent to mammal predators. Usually when leaving the eggs in normal circumstances, she covers them just with the down to keep them warm while she is away.

I looked at the oystercatchers' nest and was surprised to find all three eggs gone and no sign of the shells. While I stood there a little black-and-white chick with a short, orangey bill moved out of the grasses, as if posing for a photo on the rock above the nest. It is hard to find an oystercatcher chick for it's a master at hiding. I took photos of this one as it obligingly opened its beak in a wide yawn! I retreated slowly and, before leaving, took a look at one of the herring-gull nests that had contained eggs earlier. One chick, wet and bedraggled, was out of its egg, another was half out, while the third was just poking its egg tooth through. The three stages of hatching in just one picture, and the grasses, rocks and bright sea pinks framed the nest with lovely colours. I had been more than lucky on these two short visits and I did not go to the islet again that year.

When I walked over the hills for my mail next day the good luck spell continued. A letter from the American magazine described my article on the wilds as 'magnificent'. They were paying $600 for it. What a victory it seemed! At last I had been given the chance to write about what I really believed in, and the work had been appreciated more than anything I had ever written. My joy increased when I received letters from folk all over America saying I had brought the natural world to them in ways they had not thought of it before. It seemed then that indeed I might just be able to repay the wilderness for all it had done for me by sharing the experiences with others, and perhaps still earn the small income I needed to continue such a life.

In mid-July the mackerel shoals were thick in the shore waters and with greater experience it was easy to catch all I needed in a mere half hour, so I redoubled my summer custom of trying to feed the seals near their favourite haul-out islets. Rounding their rocky shores one morning I noticed that while all the other seals were hunching down like giant caterpillars and flopping heavily into the water, one youngster stayed on the rocks. It seemed to be struggling to go forward as it hunched hard, and as I drew closer and it turned to me with fear-filled eyes, I managed to take a photo of it from the moving boat. But it still stayed there. Clearly something was wrong.

I pulled the boat in beside the rock, put its bow rope under a large stone and walked over. The seal's right front flipper had become wedged in an unusual barbed projection in a crevice between the rocks. The harder he tried to get away the worse he became trapped. In fear of this two-footed monster now lumbering down on him, he snapped and tried to bite my hands.

'Come on now, Sammy Seal,' I said soothingly, dreaming the name up out of nowhere. 'We'll soon have you out of this.' By a quick movement I grabbed the loose fur of his thick neck behind his wide head, thankful that owing to the shortness of the neck he had little head manoeuvrability anyway, and then pushing his whole body down and backwards, I freed his flipper. The skin was torn badly but the limb was not broken for it was flapping away every bit as hard as the other.

At that moment I was sorely tempted to try and take Sammy home as young common seals are not hard to keep as pets if they have access to natural food and water for swimming. I still did not like the solitary life and briefly imagined the young seal and me sharing life together, playing and swimming in the crystal-clear summer waters. But my basic belief is that 'friendship' with wild animals should be based solely on giving help when needed, not possessiveness. Taking any youngster from the wild is only justified if one is absolutely sure it has been abandoned, or it is injured beyond self-help. Just then I saw a large, dark-brown seal, nearer than the others, anxiously circling the edge of the rock. It was clearly the youngster's mother. I realised too that the pup's flipper would heal as well in the mineral contents of the sea, with its natural salts and iodine, as through any amateur help of mine. So I let him go.

For a second or two he just stayed there, staring at me with total disbelief in his dewy eyes – seals have most expressive faces – then he hunched himself rapidly down to the water and swam out to his mother. It was only then I remembered my camera. I raced back but when I got it from the boat and adjusted the lens there was not a seal in sight! There was ingratitude – after that rescue they might at least have let me take some good pictures. But I came to treasure the one I did take of little Sammy on the rock for I saw him many times later.

It was a far wetter summer than the previous two and one afternoon, as I worked the garden, I heard hissing and felt as though I were being pelted by tiny stones. It was a hailstorm – in July! It was strange that all bird song ceased on these wintry days. If such song is only to attract females and mark territories, why don't the birds also sing in rain and storms?

In the bad weather I made two lobster creels with hazel wands and fish netting and set them out near the islets, but when I finally caught a small lobster with only one claw I felt so sorry for the poor creature I let it go. Mostly I caught only crabs or dogfish, small and spotted like leopards. As these scavengers were common I occasionally made a spiced kedgeree from them. Their skins were covered in hard, backward-pointing hooks and when dried made an effective substitute for sandpaper.

Gardening on other cold days, I used the thinnings from the young cabbages, lettuce and carrots as salads. I battled against slugs that razed away at the vegetables all night. But instead of chemicals I put old wall boards round the edges of the garden. The slugs hid under them all day and I just had to remove them fifty yards. It would take them days to get back and I reckoned they could rasp away at weeds en route. In frequent mists the midges were at their worst, converging in clouds as I worked outside. I found they were more quickly attracted to a man in a white shirt – that sight played a part in their locating a warm-blooded creature to bite, and I had less trouble in a black shirt. I had no wish to cover myself with costly skin-burning repellent all the time.

When the weather improved again other insects proved interesting and I made notes of sightings for the British Museum. One morning a loud whirring noise and the sight of a huge insect made me leap backwards. When it landed on the croft wall I identified it as a drone fly, its thick body just under 1¼ inches long. This huge fly is uncommon in the Highlands and it resembles a bee. Its larvae live in rotting carrion and it is said that this is the 'bee' the ancients saw emerging from the carcases of lions, giving rise to the saying 'Out of the strong came forth sweetness'. I also found a tiger beetle with bright-green elytra, an alder leaf beetle with brilliant-green thorax, and several pine bark long-horn beetles though there were no pine trees for a good mile. As I worked in the garden buff-tailed humble-bees biffed into foxgloves and honey bee workers landed on white clover flowers. Sometimes they forced their way through thick grass to get to them, sending up dust with their buzzing wings as if making up for time lost during the late cold and rain.

When blazing hot days came again they seemed more precious than before and I took advantage of them too, finding new diversions in the sea. With the cheque from the American magazine I bought a black rubber wet suit and flippers and spent hours snorkelling with a face mask, diving amid the lovely undine caverns and waving tangleweed to watch underwater life. It was a strange and magic world of colour, movement and silence, a dream-like ballet of strange new creatures, with myself the only privileged spectator. Shoals of silvery sand eels darted by only feet away, while afar the mackerel kept their distance, flashing their striped blues in the green depth of the ocean, and occasionally what looked like salmon smolts swam out from the river for their years in the sea before returning as great silvery salmon to spawn. Bright-red sea anemones waving their soft tentacles dotted the rocks, hermit crabs staggered along under the load of their secondhand homes, and orange, purple and light-brown starfish decorated the crevices of the dark boulders. From the gaps between them streamed flags of red dulse and sea thong, bright-green folds of sea lettuce, furry bladderwrack and banners of tangleweed.

I drifted in this quiet kingdom, body warm in the wet suit, moving buoyantly without effort, suspended in a blue-and-golden mirage. Great crabs prowled over the bright, white sand for food and I found if I wanted a good laugh I could have fun trying to poke them over on their backs with a bamboo cane. Then they ran sideways fast, and when flipped backwards they grasped the stick with a grip like a grapnel hook, actually denting it. I could fish for them while in the water by lowering a mussel on a weighted line near some weeds, then watch them scent the food and converge on it like packs of dogs. I teased them a while by pulling the mussel out of their grips until they really seemed to lose their tempers, clashing their pincers in rage as the mussel danced out of reach. I always rewarded them by letting the mussel go after a while. It was a fine comedy show – and laughs are scarce when one lives alone in the wilds.

One day, however, the laugh was on me and I had a great fright. I was still feeding the seals regularly and often saw Sammy, the little pup whose flipper I had freed, for he seemed to come closer than the others. One morning I decided to try to feed them when I was actually *in* the sea in my wet suit and mask, so I could watch them under water.

I caught some mackerel, then boated to the islet, dropping a few small pieces offshore. Most of the seals were in the water, Sammy's broad head in the vanguard, while those left on the rocks caterpillared their way down with hunches of their fore-limbs and thrusts of their hind flippers. I anchored the boat, walked to the far side of the islet in the black wet suit, fitted on my own black flippers, threw more fish out, then edged from a sitting position into the water with my linen bag of fish around my waist. I swam quietly a few yards out so I could see them clearly through the face mask.

The seals were transformed from clumsy slow-moving land creatures into dark torpedoes, as agile in the water as huge otters. They dived, rolled and twisted, sculling mostly with their rear flippers, sometimes using just one at a time, as they scooped my fish bits up in their mouths, complete masters of their watery world. When I threw more fish out I realised they were not very hungry for often they just shot past the fish, giving it a playful thrust and sending it spinning off in a vortex. One of the two larger seals in the dim distance was hanging perpendicularly in the water and allowing itself to sink down very slowly. It looked as if it had been shot and was sinking dead to the ocean floor far below through inky blue darkness. Although I had to lift my head constantly above the surface for air, some of the seals never came up for a breath at all. Seals can stay under water for over thirty minutes for they have 50% more blood than a land mammal the same size and it is specially adapted for storing oxygen and tolerating carbon dioxide.

As I kept watching, Sammy's light-grey and brown form came really close, his large dark eyes staring intently into my face mask. I could see the tightly shut Chinese slits of his nostrils and his long white whiskers. As I turned on my side to reach into the bag for the last of the fish, my face mask shipped some water and I couldn't see any more. I had to make fast strokes with the flippers to keep my head well above water while I pulled the mask out slightly, let the water out and set it back into position. While I did this and sank back into the water I was aware of a sudden disturbance and to my horror saw one of the larger mottled black seals, clearly a bull of about four hundred pounds, was boring towards me from the outer edge of the circle of seals at great speed.

There was no mistaking his intentions. I turned, thrashing the water frantically, and just managed to get to the islet again and scramble up sideways, hampered by the flippers, as he turned back with a swirl. At first I thought he would try to come after me on land too and wrenched to get the flippers off so I could run for the boat, but it seemed he knew he was at a disadvantage there. He satisfied himself with a somersault, smashing his rear flippers on the surface, then dived and came up among the others again. It was a chastening experience and I climbed into the boat with my heart pounding. Clearly my flailing movements while clearing my mask had enraged him. Perhaps, as I was clad head-to-toe in the black rubber suit, he had thought me to be a seal too, or at least something that should be driven away.

I thought that was the end of it, but as I rowed home there was a commotion on the surface a few yards from the boat, and the bull seal shot half out of the water like a porpoise, somersaulted and again hit the water with a great splash of his rear flippers before boiling down again. It was the last time I tried to watch the seals while actually *in* the water with them myself.

I realised a few days later I had chosen the worst possible time too, for as I rowed past the islet on another fishing trip I looked through the field glass and saw one of the two bulls rolling and turning in close body contact with a brown female. He seemed to be trying to bite her neck with comparatively gentle affection. It was obviously the mating season. But by mid-August the two bulls seemed to have gone elsewhere for I did not see them again that year.

One cloudless morning with a slight westerly breeze I went in the boat to fetch some of the cabin plywood for the croft walls. As I headed in I saw two otters swimming in the sandy lagoon at Shoe Bay. Having no camera with me, I made a mental note to go back in a few days' time. When I had the plywood sheets in the boat, the engine, which had been giving increasing trouble, flatly refused to start, and as I'd just polyurethaned the oars and left them in the croft to dry, I couldn't row. I decided to try to sail home. I stood in the boat, held up a sheet of the eight-foot-by-four plywood and used myself as the mast. By changing both its angle and my position in the boat, I managed to sail back round the promontories on the wind. Luckily, with such an insecure mast, no strong gusts came to send myself and sail flying into the sea. I did not realise it then but I was the world's first windsurfer.

I hauled the sheets up the slope and set them to the west wall. None would fit. The croft had sunk slightly at that end over the years and each sheet had to be planed and cut to fit its neighbour. While I was standing on a saw horse to fit the last one, it suddenly tipped over, the heavy plywood sheet crashed into the burn, and only a last-second, mad, scrabbling hop, step and a jump across the deep rocky pool saved me from a broken leg. As I lay on my back the male and female buzzard sailed overhead, the female with her beak wide open, yet not making her normal '*Keeyoo*' cry. I had the distinct impression she was laughing at me.

On my next supply trip some of my mainland acquaintances seemed to be enjoying a secret too, for everyone was most affable, greeting me with smiles. The postman stopped me on the road with 'How's life over in Paradise?' The postmistress handed me a great parcel of letters saying, 'Here's all your fan mail!' *Fan* mail? Me? The story about my search for a wilderness paradise had finally appeared in the popular magazine and, while no address had been published, 40 readers had written to me that first week. This total went up to 367 and I felt it my duty to answer each person who had taken the trouble to set pen to paper.

The letters came from nature lovers in big towns and small villages all over the land. I was astonished by the kindness of total strangers for I was sent cakes, sweaters, personal accounts, poems about nature, and recipes galore. I must confess, as I replied, I entertained the naïve hope that I might thus meet a lady with whom to share the wilderness life. I could see now that if I was alone in my mid-forties I had no-one to blame but myself, for my former fast-travelling life had been one of foolishly rejected opportunities rather than a lack of choice. All the same, I still clung to some personal illusions. Possibly some of us, the romantics or the idealists who still believe man is perfectible, or even just the lonely, hold to illusions because without them they feel stripped of hope. Perhaps if we could really see life in all its stark reality we might instead go mad. Above all, the massive correspondence helped assure me that writing about the last wild places and man's influence and responsibility to them was, for me, the right work to do.

After the first morning pounding out replies, I took the camera and walked to Shoe Bay, hoping to lie in wait for the otters. I was surprised to find some campers had been given permission to pitch their tents by the only south-facing, sandy bay and lagoon on the island which was also the otters' main playground. So that was the end of any otter-watching for a while.

On my next shopping trip Dugald Cameron, my Highland butcher, greeted me with a wide grin. 'Are you still looking for a bird?'

By now I was a bit embarrassed by the leg-pulling after the article and such English slang words as 'bird' for girl seemed out of keeping with his usual demeanour. 'No,' I said defensively, biting the wrong bait. 'I'm not looking for birds. If I was I wouldn't be living way out there.'

He laughed. 'No, you said a long time back that if anyone found a young or injured eagle or buzzard to let you know. Well, the new mechanic has a young buzzard the Forestry Commission lads found fallen from its nest. He's living in a caravan, and is working all day. He might let you have it.'

A young buzzard? It had been one of my ambitions to rear a hawk, falcon or buzzard ever since I'd come to Scotland. The memories of the birds I had kept as a boy came back – a crow, wood-pigeon, jays, magpie, owls, kestrel, and an injured female sparrowhawk. One of the favourite books on my shelves was Tim White's *The Goshawk*. If there was the faintest chance of buying the bird, I wanted it. Thanking Dugald, I dashed round to the wooden shed where motor engineer David Sturrock was starting a new and badly needed garage business in the area – the hard way.

I can't remember what I said but I talked ten to the dozen about rearing birds, how I lived alone on the island, had an empty room overlooking the loch, and a few minutes later Dave stopped work and we drove round to his caravan. The bird was crouching under a cage made from a vegetable rack, trying to avoid the playful attentions of a golden retriever and a young cairn terrier, his feathers wet from standing in his drinking water. But fear seemed not to be in his make-up for, ruffled though he was, he glared in baleful indignation at the excited dogs, ready to fight for his life if they went too far.

Shooing the dogs outside, Dave lifted the rack. Immediately the bird leaped and flew to the safety of a curtain rail, keeping a precarious balance by fanning his beautifully fawn barred wings, and staring down with intent curiosity at this new intruder. He seemed impossibly small for a buzzard, less than a foot long and despite his similar rufous colouring was far too slimly built. I only knew I wanted to look after this lovely bird whose yellow eyes now glared with luminous ferocity into mine.

Slowly I reached up and tickled the bird's underbelly. He quickly stepped on to my fingers, clinging hard with sharp, yellow talons.

'It looks like he's yours,' laughed Dave.

'But how much do you want for him?' I asked, thinking of my near-zero bank account. Dave thought a brief second then shook his head.

'Ah, he cost me nothing at all, so you can have him for nothing too! But I warn you, he eats like a horse. I've been shooting rabbits for him so you'd better take some of these carcases with you. He loves them.'

Astonished at his generosity, I thanked him profusely and with a jubilant heart carried the bird to my Land Rover. During the jolting two miles back to the little pier he remained perched, struggling to keep his balance, on a board behind my driving seat, and seemed perplexed at the rapidly passing scenery.

But the moment I pulled up he dashed and flailed away at the windscreen. My instinct was to let him free, but I dared not. He was far too young and the white down still clinging to his plumage told me he was probably not yet five weeks old. He would have been easy prey for any passing fox, wildcat, stoat or even a group of hooded crows.

As his wings whirred, spraying me with droplets, and because I still thought he *might* be a buzzard, I knew what I would call him – Buzz. I calmed him down, tied a soft gymshoe lace round one of his legs, then put him into the safety of a linen shopping bag. I loaded my supplies into the boat, cast off, started the outboard and set out on the two-and-a-half-mile journey to my cottage on the Atlantic end of the island.

11

Bird on My Hand

It was one of those calm and balmy summer evenings when the air shimmers up from the sea without any wind, and after a quarter mile I let Buzz out on to my hand. He showed no panic but regarded this new world with curious circular motions of his head.

Already he seemed happier. The air, the light, the glimmer from the sea and the trees, rocks and sky, all these were his natural element. At last his eyes could focus on a distant horizon and tattered though he was, he responded. His jet black pupils contracted and expanded as he turned his wedge-shaped head and stared with piercing marigold eyes, taking in everything around him.

As we approached the islet near my bay he suddenly stood erect, flapped his wings and cried '*kee kee kee*', perhaps calling for his parents. As I looked down his little throat and cruel beak, I rejoiced at the sound for now I began to realise what he was. The warm breeze had dried him, revealing the full splendour of his plumage. His back and shoulders were dusty chocolate brown with slight fawn edges to the feathers and wisps of white fledgling down still clung in places. His chest was white and crossed with rich copper bars. His tail feathers, already long though still encased in an inch of wax sheath, were a bright russet, striped with thin bars of dark brown.

Everything, the short rounded wings, the long tail, the brilliant yellow iris, pointed to Buzz being a hawk. Falcons have long stiff scythe-like wings and short tails, specially adapted to hunting down flying birds and small running mammals in open country. They also have dark eyes and notched beaks. As if to remove any lingering doubt, Buzz bobbed his head, lifted his tail and ejected a slashing white but odourless mute into the boat. I was delighted by this final clue – falcons' droppings fall straight down. Buzz was a sparrowhawk.

The injured female sparrowhawk I had reared as a youth in the Sussex countryside had died. Perhaps I could now redress the balance, though I soon realised the enormity of the task. As any falconer knows, the sparrowhawk is the hardest to train, especially the smaller male known as the 'musket'. It is delicate, has an intractable nature and must be treated like a spoilt child. If its metabolism sinks too low through poor diet or being overflown, it can develop fits which result in apoplexy, heart failure, and sudden death.

At this sombre thought I looked down. Too late. Buzz ducked, ejected another mute and leaped. My grasp on the gymshoe lace was too slight and he went beating across the sea for the land. My god, I've lost him already, I thought as he flapped with rapidly weakening wings, ragged and tatterdemalion like an old crow in a storm, and fell into the sea thirty yards off shore.

Scared he would drown, I turned the engine up to full speed, only to have it cut out to one cylinder. Leaping for the oars, I rowed madly to where he was floundering. Surprisingly, he seemed to be making progress, flailing his legs like a duck and humping himself forward with his wings as I had once seen a bald eagle do in Canada. Afraid to hit him with the boat, I eased near, scooped him out by hand and put him back into the bag, a scraggy shivering object by now. As I carried him up to the croft he shook and rattled his wet feathers as if nothing had happened. He looked carefully at everything – the row of ash trees, the gurgling burn, the clumps of bracken and the harsh granite boulders – as if memorizing it all.

Inside my large spare room I quickly made a perch by the window from a curved ash twig so that it overlooked the supernal view of trees, islets and sea. He seemed to like it, intently watching birds, butterflies and even moving leaves. Below the perch I made a platform for his water and where he could tear up his prey and meat. In order to be able to watch him when he was unaware of my presence, I sneakily drilled a small hole in the door at crouching height and left him. Half an hour later, shoes off, I stole back. He was now preening himself. Arching his neck high, he thrust his hooked beak into his top breast feathers, then wiggled it down towards his legs. Once they were all properly fluffed out, he twisted his head over his back and combed the secondary wing feathers out with little beak quibbles. His primaries seemed too short, barely two inches long, and at first I thought they would not grow. He smoothed these by stretching each wing out like a fan and raking the feathers with his talons. When he flirted his tail out and rattled it a few times I saw one of the feathers had broken off during the fall into the sea. His careful toilet now seemed complete.

Only then did he look at the rabbit haunch I had left. He glared at it once, relaxed, closed his eyes for a doze, then as if doing a 'double take', opened them wide again, looked intently at the haunch and crept down to the platform like a little parrot towards his prey. Although my back was aching, it was so comical to watch I didn't move. From six inches away he took a sudden leap, struck the haunch a terrific blow with his left foot talons – presumably to kill it – stamped on it twice with the other, then tore it apart with surprising strength and swallowed the torn morsels. Dave Sturrock had been right – he *was* feeding well.

I knew little about training a hawk except that to 'man' one successfully, as falconers do, it had to be trained to sit on the gloved fist or wrist, even learn to feed there. When outside and flying free, it had to be trained to return to a swung lure of food it recognized even after making a kill of its normal prey. All this takes enormous patience and one technique is to sit up with the hawk for hours at night until man and bird are both exhausted but it finally accepts the fist so naturally it will even roost there.

For the first two nights I tried the first part of this training. Making a special portable perch, I brought him into my room and installed him on my desk. Then I gently stroked his breast with a feather, talking softly to him. But I had to eat too, and when I did so, he did not sleep or relax at all but crouched on the perch as if about to spring, intently watching every movement of my knife and fork, an expression of outraged query in his eyes. I realised then how the phrase 'Watch like a hawk' evolved. His gaze was so penetrating and disconcerting he made me feel foolish. I looked at my knife and fork as if for the first time. What ridiculous implements they *were!*

Shortly after I had put him back in his own room I heard a slight scuffle. I opened the door to find he was not on his perch but on some sacks on the floor. Used to a nest at night, the floor of the cage in which he had been kept in the caravan, he naturally sought something flat to sleep on. This habit had to be broken, for if he ever did it outside, a fox, otter, wildcat or stoat could get him. When I went back in with the paraffin lamp he flew about the room, refused to land on his perch but settled for an old deer antler on the shelf above the hearth. This eventually became his favourite night perch, and before going to sleep he always drew one foot upwards, puffed out his chest feathers, sank the foot into them then smoothed them down again, all snug and warm around his toes. At least one foot kept warm that way.

Next morning I crept up to the peephole just in time to see him fly from the antler to the window perch, look at the water tin as if seeing it for the first time, then drink copiously, lowering his beak then tipping it up filled with water and swallowing it down. When I went in he glared at me as if he had never seen a human before, a way of looking right through you. Again I was impressed by his beauty, the penetrating focus of his eyes and his arrogant independence. He flew round the room with whirring wings and as I stood hopefully with outstretched gloved arm he suddenly landed on it as if he had been trained for years.

That night, when I brought him into my room, again he glared at my eating methods. When I lit a match for my pipe, he ducked, sent a mute across the floor, then flew wildly around, landed on some saucepan handles and sent them clattering to the floor.

It was then I felt I was wrong. I knew I did not really want to *tame* him, although I was sure it was possible, and I had plenty of time. But such a creature could never be a pet, even if I could get the necessary licence to keep him legally. He would never know feelings of affection, could pay no homage to a human owner. I suddenly realised I had no wish to 'man' him in the falconer's sense either, to train him to fly at wild prey, kill to satisfy any hunting instincts I might have, then return to the hand. To achieve this the bird has to believe your will is its will for, like a cat, it remains fiercely independent. But he was not born to be my indoctrinated slave. No, instead I would try and train him to hunt for *himself*. Once I was sure he could do this, and that he would return to a certain place outdoors for food if he failed to catch his natural prey, I would release him. I would help him only to fulfil the destiny for which he was born – to be free.

I caught him again, put him back on the deer antler perch in the far room and retired to think it all out. The sparrowhawk is now comparatively rare in Britain and apart from the extremely rare goshawk, which infrequently visits eastern coasts, it is our only true hawk. It is nearly extinct in twelve counties where once it was common. Years of persecution by gamekeepers, poisoning by toxic pesticides building up in the bodies of its prey, and the reduction of its best habitat – varied broadleaf and conifer woodland – took place before it was given complete legal protection in 1966. With the reduction in the use of pesticides, it has slowly fought back to re-colonize many parts of Britain from its strongholds in the west. Clearly Buzz was the son of fine pioneering Argyll stock!

He would never soar effortlessly over the mountains like a golden eagle, or quarter the ground with the incredible

long-distance eyesight of a buzzard, which is eight times finer
than a man's. For sheer speed the falcons would outfly him. He
would never match the hurtling downward 'stoop' of the rare
peregrine which hits with such shattering impact it can decapi-
tate a flying pigeon, nor the flashing flight of the scarce hobby
which can fly down a swift, nor emulate the hovering abilities of
the common kestrel as it balances in the winds above the fields.
Yet, if I succeeded, he would become the dashing Sir Lancelot of
the wood and glen. He would learn to dart, arrow-like, along
hedgerows and woodland glades, spy a finch on some outstand-
ing twig and swoop to pluck it off with one foot and unerring
skill. He would be fast enough to snatch a swallow flying from a
barn, skilful enough to chase into a bush after his prey without
suffering mortal damage. He would learn timing fine enough to
pluck a young water bird from a loch surface without a ripple,
and to use his long tail for the agile twists and turns in dense
woodland of which no other bird is capable.

Next day Buzz's education began. With his primaries still so
short I had first to help him learn to fly. I made a pair of light
jesses for his legs from the soft tongue of an old shoe. These
took an hour to fix on because he hated being touched. So that
he could fly freely I tied a 400-yard length of thin nylon fishing
line from a spinning reel to a tiny swivel on the jesses. This idea
worked wonderfully because the line could snake off the reel
without any friction. I knew one must never pull a hawk or
falcon up in flight during training, but if he were about to tangle
up in trees or round high rocks I could try to slow him down
with a light finger touch so he would turn away. This line was
essential in case he escaped to freedom before he had learned to
hunt.

On the first day he managed two 100-yard flights. On the
second I allowed him three of 150 yards each. On his initial flights
he flapped unstably like an old rook, and I wondered if his lost
tail feather could affect him that much. But it seemed due more
to hesitancy because of the newness of the area and the sheer

open space around him. By the fourth day he was flying with more determination, deciding where to go before taking off.

There were many problems, partly because man and hawk were learning together. Sparrowhawks hunt mainly in or near woodlands, and however long his flights he usually made for the ash trees along the burn. When he took off again at my approach, I had to try and throw the reel over the bough he had perched on and catch it again so he could keep flying. This didn't always work and occasionally I had to slow him down so he came to earth gently before the line became too entangled. By the ninth day he could cover three hundred yards at a time and was varying his flight by an act of will. He would fly a little, glide, decide to go elsewhere, hover briefly, then set off in the new direction.

Later, in true hawk fashion, he liked to use the ash trees as cover for his flights and dash through them, drop down low again to the slope of the hill, skirt rocks and bracken clumps, swoop up to avoid some scrub oaks, then disappear at the full length of the line. By this time the line would have passed through the branches of maybe four or five trees and it was a laborious process to get him back. I had to follow the line through the trees until it led to him. Then I would break it a yard from him, tie him to some flexible bracken, walk back, reel in the loose line, and return to him again and tie him back on to the reel.

Oddly, after each long flight, Buzz was quite wild and obdurate. At my approach he reared up with beak agape and wings extended as if I were some would-be attacker he had never seen before. But he never tried to slash me with his beak or talons. Nor, although he'd never worn them before, did he ever try to tug or undo his jesses with his beak, which I thought extraordinary. I made it a rule never to get angry with him, but as he kept exasperatingly hopping on and off my glove after his flights like some little keep-fit maniac I occasionally muttered cuss words. Strangely, it was always after such little scraps that he was most hungry.

Buzz had some odd quirks. He reacted strongly to colour. While he *would* perch on my bare hand, he absolutely refused a grey or blue glove and was finally happy with an old tattered brown one. He examined the contrast between my trousers and shirt with intent scrutiny. He did not like white shirts or if, being warm, I went in naked from the waist up. I tried out all my shirts and found he was happiest with a black short-sleeved one.

At first he constantly leaped or 'bated' from the glove the moment I opened the outer door, when we were too near the ash trees and the croft walls, and for a moment dangled humiliatingly by his legs on a yard of line. After endless coaxings back on to the glove he finally learned not to fly until I either held my hand high or had transferred him to a special ash perch outside my window where I could keep an eye on him.

This perch was most important in Buzz's young life. After his flights I always left him on it, the open reel lying below. I fed and watered him from it during the day and he soon realised he could fly to and from this perch at will. My idea was that he would always regard it as home when he went free and would return there for food if he failed to catch natural prey and was hungry.

Feeding Buzz proved a big problem. Although he ate rabbit flesh it was not his natural food, being equivalent to what falconers call 'washed meat' and not nourishing enough for a small hawk's high metabolism. A sparrowhawk's natural prey is small birds, some mice and voles, and a few beetles and small invertebrates. Even so I still needed rabbit for I had to wean him off it gradually. Lean red beef was a good food, but again it was unnatural and lacked the fur and small feathers the hawk's digestion needs. On the fourth day I ran out of both beef and rabbit. As I had been plagued by mice for years, waking me at nights and fouling my stores and once my bed, and because they were common and the hawk was rarer, I felt a few would have to be sacrificed to keep Buzz alive.

I had no mousetraps, and that night put some cheese on wire on one edge of a bucket of water and hinged a small weighted see-saw of thin wood opposite. When the mice ran up a sloping box and along it to get the cheese, they promptly dropped into the water and drowned. Next morning I had three mice – enough to keep Buzz going until I obtained more rabbit and beef.

That day the inlet valve on my boat engine came out and I broke it trying to put it back again at sea. For the next ten days until Reggie Rotheroe, a first-class engineer as well as inventor, repaired it I had to row to and from the pier. Luckily I found a newly squashed rabbit on the road to the store. After scaring away a couple of hoodies, I prised up the rabbit and together with some lean beef from the butcher's I rowed back.

The hardest part now was to teach Buzz to hunt for himself, and to do this I decided to make him *work* for most of his meals. I experimented endlessly, trying to arouse his hunting instincts. I tied a dead mouse by its front paws to some nearly invisible line and, while Buzz was dozing on one leg on his outside perch, I dropped it near him. I walked about twenty yards away and, for a minute or two, pretended to ignore him. Then I gave a little jerk on the line.

The mouse jumped realistically through the grass. Immediately Buzz went into a crouch, glaring down at the mouse as if about to pounce. Then, to my great chagrin, he looked all the way along the line to me and his expression said clearly, 'What is that damn fool up to now!'

It was not until I had put him on half rations and partially buried the line in the grass on the fourth day of this new training that the idea worked. I had caught no mice that night and was using a fur covered rabbit leg instead. Hiding behind the big rock, I gave little tugs. Buzz immediately dropped into a crouch and, when he saw his prey getting away, he launched himself, a mad stare on his face. Like an arrow he came, hit the moving leg first time, struck it two lightning blows with his talons, one-footed like a cat, then with his wings mantled over the leg and both feet on it, began to feed. He plucked out the fluff most professionally, tossing it away into the air, then started to eat the raw meat.

As I walked past with the line he ignored me completely, as if intent on his first real 'captive'. I left him for a while, then got him back on his perch by dragging the rabbit leg away from him. He stared with disbelief and half-ran, half-flew, pouncing repeatedly on his prey, glaring with rage, striking it again and again with his flashing scimitars and mantling over it each time it came to rest until he thought he had finally conquered.

Every day I varied this routine, with mice when I caught them. Sometimes I wound the line right round the cottage, using the smooth guttering downpipes to reduce friction on the corners, and came back behind Buzz tugging the mouse. He immediately went after it and it was great fun hauling it all the way round the cottage and then seeing Buzz coming round the last corner towards me, still flailing away at it, striking out with his claws like a little boxer.

When he finally carried his prey up to a branch in the ash trees to eat – thus choosing a plucking post all sparrowhawks like to use – I felt we were really making progress.

12

Symbol of Freedom

Now that Buzz had learned to catch running mice on the ground I had to help him learn the far more difficult skill of catching his principal prey in the wild – small flying birds. At first I tied some small feathers and rabbit fur round a piece of steak and swung it towards him from all angles. At first he was alarmed, dodging to and fro as the 'bird' came near. After flying him hard on the second day, he followed every movement with glaring eyes as the bait swung towards him. Then he shot out one set of talons and clung determinedly to it while still holding the perch with the other. As I tugged so he resisted every attempt of his captive to 'escape', pinning it down with great strength. Finally I left him to it and as he ate he swallowed some of the feathers his system needed.

I soon realised that this was a poor method. Birds would hardly fly within a few inches of his face in the wild. Indeed, one day the tame chaffinch Little Fat Sergeant, who had already helped to raise a brood of youngsters, flew by and landed on the garden fence. Buzz, who always watched intently every moving thing around him – blowing leaves, butterflies or soaring dragon-flies – instantly dropped into a crouch. The chaffinch froze, as if mesmerized by the raptor's eye. This seemed to be a protective device on Sergeant's part, for when he did not move again, Buzz lost interest, and the chaffinch quickly flew away.

A better way to teach Buzz to fly after birds came in a sad way. While walking down to the south shore I found a wheatear fluttering and hobbling in the grass, too weak to fly. Its legs were entangled in sheep's wool so that it could not hop or walk, and had been unable to feed. I brought it back to the croft, placed it on its back on a cushion – this seems to hypnotise small birds so they don't try to fly away – and removed the wool with sewing needles. When I fed a worm to the bird it quickly spat it out. That night I caught insects by shining a torch on to a white sheet and fed it those instead. Next day it had perked up enough to flutter at the window so I let it go. It stayed nearby and dipped and bobbed, uttering a little squeak that sounded like two stones being struck together. At dusk it hid between two rocks. Fearing the little wheatear would be taken by a predator, I brought it back into the croft and fed it, letting it out again next day. A few days later I found it dead on the path.

As I looked at its pitifully thin body, its bright blue, black, white and chestnut feathers, I realised its death was a blessing in disguise for I could use the body to teach Buzz to fly at birds. I rigged up a 'flying run' with nylon line from the roof to Buzz's plucking post branch on the ash tree so that it passed just above his daytime perch. I tied steak to the wheatear's breast, braced its wings open with a thin wire frame and made two little wire loops on its back. Then I tied my thinnest fishing line to it, mounted it on the runway and sent it flying down. By jerking the line slightly I gave it a spasmodic realistic flight.

Each time it went past Buzz saw it and crouched. On the fifth run he leaped after it, looped up, missed but pursued his quarry, finally catching it a yard from the ash tree. At first he clung to it, flapping, for he couldn't take it off the line. Finally, with one foot on the plucking post and one on the bird, he got his meal. Once again I varied the approach, setting the runway from a pole, from rocks and small trees near the croft, but always ending at the plucking post. I also collected feathers and tried to imitate the sparrows, finches, pipits, pigeons and starlings which are among the hawk's main prey.

Buzz developed an odd liking for bathing in the burn. Every sunny afternoon at about 2.10 p.m. – I knew the time because it was just after the Radio 2 weather forecast – he flew from his perch to a rocky pool in the burn some thirty yards away. Then, constantly glaring at the sky to make sure he was safe, he waded out, up to his middle, like a duck, fanned out his tail, crouched down and shuffled water over himself with half-folded wings. So much for the belief that hawks hate to get their wings wet! Then he flew back to his perch and sat in the warm wind with all his feathers puffed out, looking like a partridge, until he was dry enough for preening.

Of course, there were a few mishaps during Buzz's training. Once I left him on a rock two hundred yards behind the croft so that he would become used to being alone. An hour later, as I wrote at my desk, I heard his high '*kee kee kee*', traced his line and found him hanging upside down from the ash tree. He was scared but also indignant at the humiliation. Unfortunately he had broken a second tail feather – at the root, so I could not mend it. But I managed to straighten out two others that were bent by the old falconer's trick of immersing his tail in hot water. He now had a gap in the centre of his tail, and I would have to be careful he lost no more. With ten good feathers left I hoped he would not be too bothered by this slight lessening of his sail area.

One morning, when I had left him dozing on his perch after a meal, I suddenly heard a raucous squawking and looked out. Buzz was gone. I felt a stab of fear as I remembered how hawks are sometimes mobbed by crows and rushed outside. In a dell in the glen to the south-west of the croft, three hooded crows were leaping violently up and down. I hastened there and peered through the tall bracken. Buzz was at bay, his beak agape with rage and resentment, but there was little of fear on his face. Two of the crows were darting at him from above without going so far as to touch him, forcing his head down, while on the ground another hoodie was launching attacks with its spear of a beak. Each time it went close a flailing stroke from Buzz's rapier claws drove it back, though his movements were hampered because the line from the reel was entangled in the bracken above him. While not actually suspended, he was on short tether and the cunning hoodies knew it. But he fought the larger birds off with talons and swift snaps of his wings, sideways, as a swan strikes for maximum power.

I yelled and leaped up, scaring the hoodies away, but if I expected Buzz to be grateful for my timely appearance I was wrong. His blood was up, his feathers shrunk close to his body, and for the first time I saw his long shank feathers as he stood as high as he could to be fast and light on his feet. He reared up and tried to get away from me too, flapping his growing but still stunted wings, pathetically small but still full of fight. At that moment any living creature would have angered him.

If I had not been at hand that morning, handicapped as he was by the line, he would have been killed. Clearly I could not keep him a prisoner for ever. The longer he became used to me, the harder he would find it to fend for himself in the wilds. Now too, the first winds and rain of early autumn were starting. I had taught him all I could as a mere human and the time was near for his final freedom.

Slowly stroking his belly feathers and making the soothing noises I always used to get him on to my hand, I carried him back to the croft and his indoor perch. Later, through the peephole, I saw him sleeping fitfully, all his feathers puffed out and swaying slightly, his transparent third eyelids, the nictitating membranes, blinking under the main ones. He was roosting abnormally, on both legs instead of one, and his mutes were now an ominous light-green colour. Both are danger signs. Had the shock been too much for his highly strung system?

Then I noticed his water tin on the floor. He must have knocked it over. Quickly I filled it and held it out to him, but he took no notice, I nudged his beak into it and tilted up the tin. Instantly he hopped on to my hand and began for the first time there to drink copiously, lowering his bill and tipping it up full of water and swallowing it down, his legs trembling against my fingers. In the drama with the crows I had forgotten one of the three most important rules when dealing with hawks – always check they are freshly watered, clear their mutes away for they can cause bacterial infection, and never leave uneaten food near them.

That afternoon, while Buzz slept, I made the long sea and road journey to the butchers for some raw fillet steak, determined that he would not die. As I came back up the steep path I was astonished to see a young sparrowhawk fly past the croft, pause over the garden, then head out over the ridges to the east. Had Buzz somehow escaped? I dashed into the croft, but he was there, perched on the deer antler on the shelf above the hearth. It seemed an extraordinary coincidence and I wondered if it could possibly have been one of Buzz's own fellow fledglings. By nightfall, after wolfing down a full quarter pound of the steak disguised in rabbit fur, he seemed well recovered.

I gave him a day's rest, then trained him intensively for a few last days. By this time the sheaths on his ten tail feathers had almost grown out and he stood a full 12¼ inches on the launching pad. He was flying really well now, darting through the trees and gliding down to the shore, looking like a true sparrowhawk with the sun gleaming from the sea upon his burnished pinions. It seemed that flapping against the slight pull of the line had developed his wing muscles, and I hoped the slight tail gap would not prove too much of a handicap.

Now I left him alone for longer periods, and after his bathes in the burn he rested more often in the ash trees than on his perch. Each time I approached him he seemed a little wilder, his feathers sleeking down as I drew near. When he was hungry he naturally flew back to the perch and looked for the food he had doubtless seen me drop. He still attacked my pulled mice or simulated prey with great gusto, holding them in the grass with amazing strength against my slight tugs.

The day I let Buzz go was fine, warm and cloudy, the kind of day he preferred to bright sunlight or rain. He refused to stand still for his jesses to be removed and flew round and round the room, landing on the deer antler, the window perch, the floor

sacks, the firewood pile – the games we had played so often
before. Finally the jesses were off and I took him outside and
lifted my hand high for the last time.

Away he went, shooting through the gap in the ash trees, twisting, turning with half-closed wings like a huge swallow, and
landed in a sessile oak tree three hundred yards away. Two hours
later, as I typed at my desk, he was still there, the jesses off at last,
casually preening his tail feathers through his bill, as if he knew
he was now completely free and there was no hurry to go
anywhere. When I looked up later he had gone – and so had a
steak 'mouse' I had put near the perch. I felt sorry not to have
seen him take it.

As I swept out his room, cleaned up the newspaper sheets with
his mutes, I saw a few wispy down feathers, symbols of his childhood, still clinging to the ash perch by the window from which he
had surveyed the rugged landscape and shimmering ocean that
were now his own world, and felt oddly lonely. My mind seemed
filled with the memory of his piercing eyes and whirring wings as
he'd flown round me before allowing me to attach his creance line
for the training flights each day. Yet I was happy that he had
survived and was free.

For three days there was no sign of him apart from a brief
glimpse of a hawk's form passing swiftly across the garden. On
the fourth there was a flutter of wings outside, a shadow crossed
the window and I looked out to see Buzz on his perch, tearing
hungrily at the food I had left out. I did not go outside for I didn't
want to frighten him.

On my next supply trip I heard that one of the summer visitors had been given permission to shoot the big stag that was
known to be on the island. I feared then that Sebastian, the
friendly stag I had been watching for more than two years, might
become the recipient of a rifle bullet for he still occasionally came
round the croft at dawn. But the stag-stalking season was now
open, it was not my land, and there was nothing I could do to
prevent the slaughter. Two days later I found a dead herring-gull

floating in my bay, something I had not seen before in two and a half years of walking round the island. Its body was shattered, most of the backbone missing. It looked as if it had been shot with a rifle. I felt a pang of fear for Buzz's life and resolved to try and catch him again next time he returned and keep him until I felt the danger was over. That afternoon, while I was painting the new west wall of the cottage with an undercoat of light grey, a piece of rabbit disappeared from below the perch – yet I had seen and heard nothing.

On the morning of his ninth day of freedom I was delighted when Buzz returned again. He flew to his perch, dived at some meat scraps I had put out and carried them to his old post on the ash tree branch. He looked more mature now, somewhat lighter in colour. I went out and found myself staring into the wary eyes of a stranger, and he flew away again.

Next morning the first rain for nearly a week began to fall. As the wind increased I went into Buzz's room to open the window in the forlorn hope he might seek shelter and return to his window perch. Something made me look down – and there he was, lying awfully still on his back just two yards from the cottage wall, as if he had been thrown there.

I carried him indoors. His body was still warm and his gullet full of blood. In terrible grief I examined him and could find no external injury, yet I noticed odd differences. The body seemed darker, smaller, and there was a patch of white feathers behind the head I had never noticed before. The wax sheaths on the tail seemed longer than Buzz's had been on the day of his release, and there were twelve tail feathers when, after Buzz had lost two, I had counted only ten. With his continual fast movements I could possibly have miscounted earlier.

I looked at the little body on my desk, remembering how he had lain on the ground as if thrown there. Suddenly, in awful doubt and suspicion and then rage, I rushed out and raced round the croft area and the ridges in a wide circle, cursing the bracken that tore at my legs. I heard and saw no one. Could it possibly be a different sparrowhawk, I wondered. Then I recalled the young one I had seen fly past when Buzz had been in his room. No, it seemed impossible. But in those heart-breaking moments I was in no state for balanced reason or any attempt to muster precise memory. And all my photos of Buzz had yet to be developed. I *had* to find the truth of what had happened for if any evil had occurred I needed positive evidence before taking action.

In despair, I wrote a fast letter, raged across the stormy loch under a steely, raining sky, telephoned SSPCA officers in Glasgow and Oban, then drove the body down to a police station twenty miles away from which the Oban officer had promised he would remove it for autopsy.

For a suspenseful week I waited, going over and over again my weeks with Buzz. Could he have picked up some poison? Or been hit by a stone? Could something have scared him so much that he had flown blindly into the side of the house or the window? I had been sitting quietly at my desk at the time, so surely I would have heard something . . .

Then the autopsy result arrived. '. . . The most significant lesion found was a large bruise, an inch or more in diameter, situated at the shoulder at the junction of the left wing and the body . . . There were no penetrating wounds; no evidence of gunshot. Examination for infectious diseases was negative, and in the circumstances there are no grounds for suspecting poisoning. The site of the external injury is consistent with the bird having flown or dived into a hard object at speed. (Overhead wires? A fence?) The ultimate cause of death was internal haemorrhage, probably hastened by the occlusion of the trachea with blood.'

Sadly resigned now to the fact that somewhere, somehow, I had failed, I left for London on a brief but long-delayed trip, to have my photos developed at a good studio, and to pick up a new 1,000-mm telephoto lens.

It was on the day after my return that the extraordinary dénouement of my life with Buzz occurred. I was on the beach making minor fibre glass repairs to my boat when I heard little finches chittering with terror. There was a flurry of wings and out from some nearby scrub trees one of the small birds flew with a hawk after it. As it passed over my head the finch looped up backwards, the hawk swooped up after it and, as it spread its tail to make the abrupt turn, I saw with astonishment the gap in the feathers, and the shortened wings as if the primaries were still not yet fully grown.

I stared, my heart beating heavily. The hawk suddenly lost all interest in the bird and I stayed still as it flew slowly just above me. It alighted only a few feet away in a small oak on the side of the cliff. It was the way he landed, with a quick flapping of wings, unlike the silent drop off the curve like mature hawks, that also convinced me. He looked greyer now, bigger and better feathered.

Suddenly he puffed out all his feathers, gave that little contented wriggle of the neck back into his shoulders, then delicately raised his right foot into his breast feathers, smoothed them down over the foot just as Buzz always did, and *went to sleep*.

Extraordinary though it may sound, it was as if he had been sent to end my tortured doubt. No wild hawk would have done that, land and go to sleep almost within my reach. As if in a dream I stood stupefied for a full minute, then said 'Buzz' loudly. Instantly he became alert, smoothed down his feathers and, as silently as an owl, flew close to my head before he was lost in the trees to the west.

I walked back to the croft with an indescribable feeling of relief. The hawk that had died had been only an extraordinary coincidence. Perhaps it had believed the newly painted light grey wall to be part of the sky. Later, as I consulted my diaries, I found items I had forgotten. Apart from the hawk that had crashed through Iain's window months before, I had also found remains of other sparrowhawks earlier in the year, one near Arean and one near Shoe Bay, both upturned on the grass with only a few

feathers left. Certainly fast-flying sparrowhawks suffer heavy
mortalities, sometimes killing themselves by flying too hard after
prey into brambles, thorn bushes and tree branches.

As I walked the woods and mountains in later months I
caught occasional glimpses of Buzz, dodging and twisting
between the trees like some swift spectre, in full control at last
of his real hunting speed. Occasionally he still came back to
snatch meat from the front patch. I hoped one day he would
find a fine mate, crying '*kek kek kek*' as he courted her through
a woodland glade and helped her build their nest in a high conifer. I hoped he would become a good sparrowhawk mate, bringing food to her and the chicks and guarding her attentively
when she stood with open wings shielding their young from sun
and rain. I liked to think the patient training was not in vain
and that at least one human had done a little to redress the careless toll taken by the species that believes itself to be the most
evolved on earth.

Some readers might think that I cared too much about a 'cruel'
hawk which, among its wide range of prey, kills small singing
birds, forgetting that many of these birds are predators in their
own way too. But in Britain the hawk is still quite rare, and far
more damage has been caused to both hawks *and* singing birds
by many aspects of man's civilized interests. No-one can pretend
that any predator deliberately goes after sick, foolish, unwary
and sub-normal birds but these are the ones it most often catches.
Thus the hawk's hunting is largely selective and is in keeping
with the primary law of the wild kingdoms, the survival of the
fittest, to keep a species fit, alert and healthy.

Above all, I loved Buzz for his great beauty, his intractable
nature and natural courage. Against a more powerful force or
being he would have died rather than give in. For me he epito-
mized what it takes to be truly free. Fear only aroused his natural
resistance, became a spur to inflexible will. Tiny though he may
have been, he was not equipped for compromise, and symbolized
what so many of us have lost.

The sparrowhawk I had laboriously carved as a boy, over a year and a half of school woodwork lessons, had been a symbol of freedom. Now, thirty years later, Buzz had become a living symbol, unifying past and present, giving the future more meaning. I learned as much from him as I was ever able to teach him, and he will live for ever in my memory.

13

Creatures Great and Small

When, in early August, I heard that a large stag had indeed been shot by a summer visitor, I felt apprehensive. Later Iain told me that, while it had been a good beast, its antlers had held only seven points. I was sure Sebastian now had a better head than that for even back in early June he had seemed to be growing a really good set. Eilean Shona was a stag sanctuary in winter, but in summer the big deer often swam between the island's north shore and the mainland, and as I hadn't seen Sebastian for some weeks I wondered if he had been killed on another estate.

Then, one misty dawn, I looked out of the window to see him standing where he had often grazed before – about two hundred yards south-east of the cottage – in a lush grassy patch by an area of high bracken. His antlers were magnificent, still sheathed in the soft velvet which he would soon start to shed. I counted ten good points. I sneaked slowly outside and tried to focus the new 1000-mm telephoto lens on Sebastian. It was a heavy 12-pound mirror lens, and focusing it was a delicate matter. As often happened in misty early morning Highland light nothing at all registered on the camera's meter – not even exposure at a 30th of a second, which would be too slow with a hand-held telephoto lens. From the huge slots made by his hoofs round the muddy perimeter of the roofed garden I could see that he had been trying to get at my vegetables during the night.

By this time the garden was in full flourish with 71 lettuces, rows of healthy carrots, and 53 transplanted cabbages where the radishes (long gone in salads) had grown, and I evolved a plan to get close to Sebastian with the less-powerful 300-mm lens which let in more light. As I picked vegetables for supper wearing gloves soaked in spruce needle juice – a trick I'd learned in Canada for reducing human scent – I laid many outer leaves in the area where he was now grazing at dawn. For three nights I left him to them, and on the last two mornings they were eaten.

About an hour before dawn on the fourth night I stole outside wearing spruce-boiled camouflage clothes and dipped boots and lay down in the thick bracken.

The wind was coming from the north-east, blowing towards me from where Sebastian normally came. I lay there for some forty minutes, totally immobile, and was just thinking I was probably wasting my time when I heard a faint cropping noise. At first I was sure it was made by sheep. I raised my bush-hatted head extremely slowly. There was Sebastian, only about eight yards away from me, much too close for the telephoto lens. With agonizing care I unscrewed it and put on the standard lens. To my astonishment he came even nearer, went down on his knees, then with a slight flump sat down right by the edge of the bracken, mere feet away. I might never have such a chance again.

Despite the bad light I opened the lens to full aperture, raised up slightly and click – one shot. Click – another. The stag turned his head towards me but showed no fear and did not leave. Another click, then he rose to his feet. I was scared then that he would come to investigate, knowing something was there but not quite what. A stag nearly always runs from man, though fatal attacks have occasionally occurred, and my heart seemed to be pounding loud enough for him to hear it. I took another picture. Then, to clear the bracken fully, I raised myself slowly to my knees. He just stared in curiosity, as if wondering what the fool who lived in the old croft, and who had never done him any harm, was doing there on his knees in the bracken! He trotted off a few paces and turned round – for another photo. Finally he walked away to the east. Screwing on the 300-mm lens, I followed him slowly, taking more photos of him. He seemed more bewildered than scared as he disappeared among the trees along the shore.

Because of the poor dawn light the pictures were not perfect, but at least they were of a wild stag close to in its own habitat, not a park or feeder animal, and conveyed the realism of those exciting moments. Unfortunately, I had blown my cover and he never came to that precise area again.

A week of gorgeous sunny weather replaced the frequent wet days of summer so far. Purple knapweed and thistle flowers which I had deliberately spared from the bracken scythe, so that bees would have their nectar, now thronged with many insect species. I was surprised to see – as in the earlier foxgloves, violets, wild parsleys, daisies, speedwell and buttercups which I had also protected – that many bluebottle flies also visited the flowers. It seems that these 'pests' (from the human point of view) also play a part in flower pollination. Butterflies too were prolific. Green-veined whites were still flying like drifting snowflakes but the flowers were now covered with tortoiseshells, small blues and browns, an occasional comma and magnificent dark-green fritil-laries which were not common in the coastal areas of the north-west Highlands.

It is often written that insects are brightly coloured as a warn-ing to predatory birds that they are not good to eat. To me, it seemed as I watched the most brilliant and fast-flying butterflies that the flashing colours as they flew, and the constant opening and shutting of their wings when feeding, were designed to help them locate each other by sight. This is the initial attraction, then scent and behaviour patterns help them to mate with the right species.

I was surprised to see one of the fritillaries persistently chas-ing after a tortoiseshell. Time and again the larger fritillary swooped after it, but if it suffered from defective vision or felt inclined to mate with the wrong species, the little tortoiseshell clearly did not. It led the fritillary a merry dance round and above the croft, before swooping back again, honing straight on to the flowers. I even got a picture of this merry chase.

I was now feeding the seals every time I went fishing, and it was clear that Sammy was far tamer than the others. When I made the usual soothing crooning sounds that seem to allay their fears and arouse their curiosity, he usually flopped into the water last. Within a minute he would be up again, only yards from the boat, his snub nose and 'bowler hat' pate far in front of the others. He was usually the first to dive for the fish I threw in – as if showing he had the most right to the easy food. I used his tameness to get several photos, but it was difficult in a rocking boat.

As he twisted and turned with bits of fish in his mouth, closing his eyes with that blissful look seals often have when playing in the water, I saw his right flipper had healed completely. There was thicker lighter-coloured tissue where the wound had been.

Through September my garden still provided me with large, succulent lettuces, and the craneflies now laid their eggs round the croft. What powerful instincts they had to perpetuate their species for they risked damage to their soft single wings and flimsy legs, bending them against the stiff grass stems to force their ovipositors deep into the soil to ensure their eggs would survive. The leather jacket larvae, well-known garden pests, I left alone as I wanted the life around me kept in a natural balance. I had seen the oystercatchers eating them, and although it's not commonly known, the adult craneflies are an important food for red grouse, supplementing their usual diet of ling, heather and cranberry shoots. On long treks round the island I seldom saw more than three or four grouse.

I was reminded of this lack of red grouse in the western Highlands when I was lent a copy of *A Hundred Years in the Highlands*, in which Osgood Mackenzie recorded the prolific wildlife and game in the region in the 1800s. The family game book for 1868 included 314 grouse, 33 black game, 49 partridges,

110 golden plover, 35 wild duck, 53 snipe and 184 hares, without mentioning the numbers of geese, teal, ptarmigan and roe deer that had also been taken. 'In other seasons I sometimes got as many as 96 partridges, 106 snipe and 95 woodcock. Now so many of these good birds and beasts are either quite extinct or on the verge of becoming so,' the book lamented.

Perhaps it does not need a scientific sleuth to guess at one of the main causes.

Mackenzie records how he collected 200 gulls' eggs from Loch Maree islands in one afternoon, how he took osprey eggs as a boy, (though 'alas, the birds have been extinct in the region for at least 65 years'), how otters were shot in their dens, how his mother had forty to fifty marten skins brought to her by keepers every year from which 'she made the most lovely sable capes and coats for her sisters and lady friends', how he once shot three whooper swans one afternoon, and how a foxhound pack killed a mother wildcat and her six kittens.

The book is a valuable historic document for it clearly shows the fatuous attitudes towards most forms of wildlife that existed in those days, the total lack of a universal view of life, the poor understanding of the roles of predator and prey species and the complicated food webs that exist in all nature. Such attitudes and arrogant assumptions are based purely on tradition which, as in many other parts of the world, is often itself based on stupidity and ignorance. Such an outlook, in the Highlands specifically, led to the near or complete extinction of such creatures as the wild-cat, the pine marten, the polecat, the white-tailed sea eagle, and the kite, and to severe decreases in the number of otters, hares, badgers, swans, kites, geese and ospreys.

Despite the unintended comedy in some parts of the book I did not feel like laughing, for it seemed to me that many of these attitudes towards non-useful wildlife still prevailed in the Highlands, as well as in the other few wild parts of Britain. But after reading it one afternoon I went on a long hike and was treated to a short comedy show.

As I emerged from the Shona forest I saw a raven chasing a sparrowhawk. At first I thought it was Buzz, but then saw it was a larger bird, a female. Suddenly she swerved down, sideways and back up, outmanoeuvring the great black bird by surprise tactics, using her longer tail to turn faster. Then she dived down upon the raven, which swerved only slightly before flying on, not at all worried by the apparent reversal of roles. It seemed it had been merely playing a game with the hawk and had had its kicks! As the hawk glided back into the woods behind me the raven seemed to confirm this by turning almost on its back twice and making a loud '*boing*' sound in its throat, like a spring twanging. It was the first time I had seen a raven fly upside down other than in spring.

That evening the air was cold round my legs as I sat at the desk. Summer was on its way out. Recalling the cold of the last two winters, I insulated the hardboard sheets on the front walls with treble thick newspaper sheets. When trying to break into London as a country reporter I had saved money by sleeping out in the big parks – and had found there was no finer cheap insulator against cold than newspapers.

In mid-September, after four rainy days, a few hours of blazing sun encouraged a new batch of insects. I counted 215 gnats and small flies on the window at night. During one hot afternoon, it seemed, millions of insects hatched, but when the heavy rains returned two days later, accompanied by cold early autumn gales from the north-east, most of them must have perished without breeding. Coming up from the shore I saw some of the weaker butterflies and small moths dying on their backs, stuck to the ground and to rocks by the wet, or washed up on the beach after being blown into the sea.

Nature is ever callous towards the individual life and allows the greatest wastage of all in the insect world. I went indoors and lit the paraffin lamp, wondering how long I could hold out *this* year without a fire. I noticed how the light was shining on to the two largest ash trees outside. A strong wind was stirring them and suddenly the broad heads of the trees looked like a man with

a beard kissing a girl, the smaller tree paler, as if a blond. His
head was on hers, a branch had become his right arm around her
neck and as they swayed he seemed to be pulling her closer. Odd
how a little illusion like that can remind one of the isolation of
the new winter ahead.

After brief showers on 21 September I was heading down to
the beach to sack some seaweed fertiliser, before it became too
heavy and wet, when I found a huge fat drone fly, dopey with the
cold, on a flag seam, as if nearing the end of its life. I saw also a
green-veined white butterfly clinging to a single purple knapweed
on the shore line. How touching that it had chosen to die clinging
to the only beautiful flower left in bloom there. A few large orb
spiders had woven their last webs, the brown females with white
crosses on their backs in the centres while smaller males waited
in the wings, tugging at the lines to signal that they were ready to
mate but were not prey.

Next day I boated out in increasing winds in the stiff-sided
rubber boat, anxious to post mail before the weekend. The engine
began misfiring, clanked loudly, and then broke down altogether.
I paddled with my hands to the shore below the forest, walked to
the big house where the Vanes lent me a 1 ½ hp engine, carried it
back to the boat and finally made it to the pier. Dave, the motor
engineer who had given me Buzz, took in my engine and lent me
a small Seagull outboard. I took the 1 ½ hp engine back to the big
house, but on the return to my bay the Seagull also broke down
in the mist and rain, and I had to paddle once again with my
hands the last mile home. A fine omen for the new winter, I
thought. As I came up the path, arms aching, the poor drone fly
was still on the flag stem. I took it into the house to dry in a
matchbox.

Two days later the mist lifted, the drizzle ceased and the sun
shone so strongly in a blue sky that it was like being back in the
perspiring days of summer. I put the torpid, unmoving drone fly
out in the sun. As the hot rays warmed it up it began to move its
head from side to side. This odd-looking, almost peering

movement of so large an insect was most disconcerting for it almost seemed to be a *thinking* animal. It began to crawl up a bracken stem, then with a sudden brief '*buzz*' it flew strongly away.

Bees had appeared again, hover flies had also emerged from hiding and zinged near my face as I sunbathed. Later I went fishing, caught some unusually late mackerel and saw the most beautiful sunset I had so far witnessed on Shona. As I walked up the slope with the fish, pulling a cabbage from the garden and slooshing it in the burn without a pause in stride, I glimpsed a blue-grey flash in front of the croft. A sparrowhawk, surely Buzz, shot up from the grassy patch by the garden. He kept his tail closed, so I could not see any gap from the lost feathers. He had been at some meat scraps hauled off the bird table by the robins, which preferred to eat from the ground if they could.

When Dave broke the bad news about my boat engine – big end worn out, crank and one piston ruined – I knew it would take time to get the parts and be a costly job. Oh well, I had a good food supply and much writing to complete. By now, after working over a quarter of a million words from notebooks, old letters and diaries, I felt the Canadian wilderness book was at last coming right.

For some time Dave and I had been discussing a possible boat trip up a long, undeveloped freshwater loch which seemed worth exploring. It was a sunny day, with three hours of daylight left. Why not *now*? We drove to the loch, hired a boat and after four miles, dodging various sandbanks, found ourselves passing long pine and larch forests. In the widest part of the loch we came to an island, where great snags of dead trees lay against each other like the huge whitened bones of prehistoric animals. The whole scene reminded me so intensely of the wild Canadian inlets I was now writing about that I half expected to see a grizzly saunter out to graze on the marshy grasses. A golden eagle sailed low overhead, black against the sky, her long broad wings upturned at the tips by the warm southerly winds. Two cormorants winged by like great dark geese to land on the branches of a dead lone spar. As we headed towards a small inlet on the south shore, high conical mountains wreathed in haze seemed to go on for ever, and as we entered the bay, the steep hillsides covered with dark spruce trees, the strange impression increased.

'It's magnificent,' I said. 'What a place to live in! It's as fine as anything I saw in Canada yet somehow more intimate.'

'Ye'd have a wee job finding a cottage available up here,' said realist Dave, who had been trying for months himself to find a place of his own and leave the caravan. 'What few exist are used once a year for stalkers' shelter or for a month or two by summer homeowners.'

As we came out of the inlet I glanced up the loch. In the distance, illumined by a shaft of sunlight, stood a tiny white cottage flanked by two woods. It was a good three miles further on, and with our little engine, the wind increasing, the sun low, we did not go to look at it. I felt sure it would not be available. On the way back we walked over a little green island that had been a Highland burial ground for centuries. There some of the great Highland chiefs had been laid to rest, including John of Islay, one of the Lords of the Isles and the first to be called Chief of the Clanranalds, who had lived at the castle the ruins of which still stood gloomy sentinel over the sea loch where I now lived. I recalled that my mother's father, John Stewart, had been born on Islay. I suddenly felt quite strange, and as we boated back in the gathering dusk I kept seeing the little cottage and the small woods in my mind.

I mentioned my feelings on journeying up the freshwater loch to Allan MacColl when next I visited his shop on the mainland. 'There is a wee cottage up there you might get,' he said, and gave me a few ideas on how to go about it. To my surprise, he was referring to the white cottage I had seen. A few days later, I contacted the landowner. To my further surprise, he said that, while it might be difficult because of forthcoming new leasehold laws in Scotland, he would think about it.

One glorious afternoon in October I saw three tortoiseshell butterflies flitting near the roof outside the croft. They were looking for shelter in which to hibernate for the winter. Even on what seemed a perfect summer day the shortage of flower nectar, the decreased amount of light, the nip in the air, told their delicate mechanisms that winter was on its way. I left the doors open, and when one finally winged into my room and lodged in the corner above the bookshelves, it seemed I would have at least one companion in the harsh months ahead.

During this short Indian summer new buttercups and daisies started flowering and humble-bees were working them along with the white bramble flowers which were always the last blooms of all. I came back from a walk at dusk to find that a wren, looking for a warm roost, had come into the croft and fallen into the bucket of water I used for cooking. It seemed near to death but was still swimming feebly with shuffling wings. I dried it out by placing it on a warm cushion, then caught some insects for it from those attracted to the window by the lamp light. Every time its beak opened when my hand went near, I popped them in with a pair of tweezers.

I let it spend the night in a hay-lined box, and even put the paraffin heater on. By morning it had recovered and was fluttering at the wire netting on the box. When I took it outside, it struggled with incredible strength for so small a bird. Like shrews, which can eat well over twice their own weight of worms and insects in a day, the wren has a high body metabolism and next to the hummingbird, perhaps, is probably the strongest bird in the world for its size.

There is a fine Gaelic story about the wren. One day in the misty past, all the birds assembled in a glen and decided that the one who could fly the highest would be their king. The golden eagle, of course, was sure he would win. But the wee wren shocked them all by challenging the mighty eagle. Up soared the great bird of prey, who cried, 'Where are you now, silly little wren?' But the wren had secretly perched on the huge bird's back and, being fresh, he flew up even higher, calling out as he went, 'Fad, fad os do cheann!' (Far, far above you!') And in that way the wren became king of all the birds.

By mid-October, although craneflies were still laying their eggs, moths and other insects no longer came to the window. Curlews were again assembling on the beaches for easier pickings than on the cold lands where they had nested and spent most of the summer. They now made curious bubbling calls when alarmed.

One night, when I went to fetch some water with my lamp in the dark, I was startled by a great grey, ghostly shape rising like a giant bat in front of me. It was Harry Heron, right by the croft. I kept watch then, and found he was flying in close at dusk and landing in the burn to catch frogs and other small water creatures. It was clear he could hunt at night and could fly straight home in almost complete darkness. Over the centuries, to escape man's presence and often direct persecution, foxes and otters slowly became more nocturnal creatures, and I wondered if this might be happening with herons too. Or perhaps he came because more small water creatures emerged at night to scavenge for their own food.

Harry was not as wary as most wild herons but he was by no means tame. Although he occasionally swooped over my moving boat with hoarse '*kraink kraink*' cries, when I tried to stalk him on foot on the beaches he took off before I was in reasonable camera range. He had also renewed his friendship with the lonely lame seagull I had often seen flying on nearby beaches. This was odd, for normally gulls and herons exist in a state of mutual suspicion. Herons have even been known to pluck young gulls from island nests and, when a heron catches a fish, nearby gulls (especially the greater black-backs) will swoop down, hoping to make it drop its prey for them to snatch.

It took several days of frustrated stalks before I could get close enough to see what was happening. The two birds often stood close together on windy rocks – with the gull apparently using

Harry's much larger body as a windbreak. They also seemed to have evolved an odd game, for sometimes the gull would leap into the air, circle and then dive over the heron's long-necked head several times, with poor Harry having to duck as the gull shot closer and closer. Yet Harry never flew away. Certainly the gull was a loner, not part of the gregarious herring-gull flock, perhaps because it had a twisted foot. Maybe this is what it and Harry had in common.

When the fine weather spell ended a feeling of real winter came. The lamp had to be lit before 6 p.m. now, rain drummed on the roof, drips from leaky gutters plink-plonked outside, new mice made nightly thumps in the wooden walls as they tried to reach the food stores, the ash trees stooped under the weight of soggy leaves, and the path filled up with slippery ooze. The bracken had turned yellow and with the nettles, now spidery brown with seedpods, it fell in swathes before the new gales. Heavy rain really hastens the look of winter upon the landscape by beating everything down and drowning the roots of small plants.

With my engine still not repaired, I had to row the five miles to and from the pier for supplies. It was a hard trip back. Against the wind, I made barely a yard with each pull of the oars. The darkness was almost pitch black, but the sea always gives off a faint luminous light at night and once you know the shape of the hills and can just make out their tops in the dimness, you can reach home safely.

14

Only Myself Left

I had hauled the boat on to its wooden cradle and was walking up the steep muddy path when I was startled out of my wits. I hadn't noticed the large stag standing on a low bank in the burn before. Suddenly it gave a loud roar when I was less than ten yards away. It was not a sound of anger but the long hoarse bellow stags make in the autumn rut, and it was the first time I'd heard it that year. I was sure it was Sebastian for in the twilight I could just make out his form, looking as big as a moose. When I neared the croft, heart thumping, I cupped my hands and roared back, and we had quite a contest until my sore throat gave out. His roars came nearer, as if he were following me. I dived inside for my torch but only glimpsed the whitish tan of his backside as he bounded away.

Next day I helped Iain MacLellan, his wife Morag and three others to gather the island's sheep for sales and winter dipping. They were short-handed. The number of men who would turn out for this hard chore on a mountainous island, involving a double boat trip and hard hill footslogging over a long day for a small standard wage, had dwindled yearly. After that day I was not surprised. We had to drive the sheep in three separate wide arcs up and down the hills, over peat bogs, through the woods and into a large stone-walled fank at the eastern end of the island. After supper at Iain's home I still had to walk the two miles back again to reach my own home. As I came near the croft in the dark, tired out, I was once more greeted by the rutting roars of Sebastian, but from further away this time.

Early next morning I tried to photograph him with his hinds. I found them almost a mile away in a deep corrie on the Atlantic edge of the island. He had only nine hinds with him. While the island was a stag sanctuary in winter, the hind population was never large, perhaps due to the competition from grazing sheep. Red deer, the largest wild mammals left in Britain, have a mainly matriarchal society. All year the hinds, their young up to three years old and new calves born in early June, keep to their own herds. The mature stags leave the hind herds at three years old and keep their own company in loose bachelor groups, usually grazing different areas from the hinds which crop closer than they do. But come the rutting season at the end of September and early in October, the stag groups break up. One by one, the big master stags first, they come down from the higher hills and look for hinds to round up into their breeding 'harems'.

Sebastian, probably a four-year-old when first he came round the croft, had carried eight points last year and bore ten this. Like all stags, he would have cast his antlers in early April. Almost immediately the new ones had begun to grow under a soft nutrient velvet skin covering through spring and summer. From mid-August the velvet would slowly peel off, helped by rubbing on trees and bushes, and his antlers would then harden. At the same time his sex glands become stimulated, his mane grows, his neck thickens and by late September Sebastian, who would have shown no interest in females for some ten months, would be ready for the rut.

As I watched, safely hidden leeward of a knoll, he walked round his harem, the hinds totally unconcerned by his presence. Now at his most aggressive as he sought to ward off any rivals that came near his 'wives', he frequently paused to roar his challenge across the corrie. Sometimes the roars tailed off lamely, as if with frustration, and sounded rather like an old man groaning loudly with toothache. These roars seemed less of a challenge than an intimidation, to warn other stags to stay away. Higher up there was another smaller, younger stag which had only three hinds and two calves with it. He gazed down at Sebastian's nine as if coveting some of them but he did not venture too far down. Twice Sebastian rushed at hinds as if to mount but both slipped away easily. Occasionally he trotted round them with short blatting roars, '*oh oh oh*,' as if he were a large horned sheepdog rounding up sheep.

The concept of the 'Monarch of the Glen' is not true as far as the hinds are concerned. Apart from mating with a stag at the precise time *they* are ready – hinds are only in season for about 24 hours in a 21-day period at this time – they take little notice of him. In the presence of danger, such as the scent of man, it is normally the leading hind, often an old yeld, who barks the alarm, and when she runs the other hinds and calves follow her, often leaving the huge amorous swain gazing after them in dismay.

Watching Sebastian and his hinds at dusk on another evening I noticed that he stood quite motionless by one hind, scenting her as if trying to tell when precisely she would be ready. She licked his neck and face with half-closed eyes. Then he trotted round behind her and mounted and served her with one convulsive heave. The hind seemed to hunch, he slipped down again, and it was all over. I wondered if red deer had evolved this swift coupling from the days when bear, wolf and lynx – and later man – were their predators and they needed to be alert at all times!

Sebastian seemed to lose interest in that particular hind and an hour later when she moved away from the group he made no effort to stop her. It seemed likely that if she came into a second season a younger or weaker stag would be able to serve her, but by that time the big stag's seed had been securely planted.

Stags – usually hard to approach in summer – are far less wary during the rut. For this reason, and because they are in peak condition after good feeding, early October is the time when stalkers come out for the stag shooting. If the sport is engaged properly, it culls the surplus animals from the hills. The hind shooting (regarded as a chore and not a revenue-raising sport) is carried out by estate staff between 21 October and 16 February, although of course, as with the stags, hind carcases are sold for meat.

To allow red deer to become too numerous on the Highland hills is poor conservation. After several mild winters their numbers burgeon – they outgrow their feeding ranges, and then control is necessary. (The deer population increased from about 180,000 in the mid-1960s to some 270,000 in 1979 and to nearly 300,000 in 1990, so that 50,000 hinds were due to be culled in the 1990/91 winter.) Man has wiped out the deer's natural predators and if he does not assume the predator role himself, the herds carry many old and runty animals prone to disease. Mortality through cold and starvation in winter is then tragically high. When grazing and browsing thus become scarce, red deer invade agricultural land and young forests, break through fences, browse the growth tips off young trees, strip bark, and inflict great damage. And not only to man's artificial commercial forests for, along with sheep and uncontrolled heather burning, they also prevent the natural regeneration of the few wild forests left and new young trees on the hills.

When Sebastian was with his hinds, he showed curiosity and aggression toward any source of disturbance, and would some-times trot closer, just as he did if I imitated his roars. Poachers can make use of these moves, just as they use powerful torches and vehicle lights at night – red deer are less afraid of vehicles than of a man on foot – to blind the animals before shooting. There are heavy penalties for such knavery. Many deer escape with fearful wounds from poor shots by poachers, just as they do from the occasional incompetent stalker, and die lingering, agonizing deaths in the hills.

During one dawn walk I came across Sebastian having a wallow in a peat bog where he seemed to have trodden the peat and mud into a blackish cream. It seemed as if he did this to rid himself of flies, keds and other pests, and he emerged looking dark and fearsome, so perhaps he also did it to look more intimidating to other stags for deer are more scared of dark objects than light. Again he just walked away, showing no sign of aggression or great fear. It surprised me that he had not been shot but perhaps he knew me to be a harmless individual. Anyway I strengthened my garden roof to keep him off the cabbages that were in top shape now.

By mid-October, with natural food harder to find, Little Fat Sergeant came back to the bird table with his family but his hen and two youngsters disappeared after two more days, probably migrating to the village area at the head of the loch. Bees were still working the brambles and a big blue *Aeshna* dragonfly was hawking along the burn for insect prey but its wings were brittle and rattled with old age.

Now the north-west winds began to blow, bringing heavier rain and churning up the seas. I lit the paraffin stove on 19 October, having lasted without it only one day more than last year. Curlews were clustered on the sandbanks, all facing the winds, a sure sign of stormy weather. Two days later Dave had my boat engine repaired and so thankful was I to get the 7-hp outboard back on to my boat that I paid the £60 bill willingly. As I returned home, no longer having to row, a flock of fieldfares with dark bodies and whirring wings flew over the boat, heading for the rich farmlands to the east. I counted 51 oystercatchers bunched miserably together on a grassy island above the main sandbank near the pier. They all faced the same way, into the gales, and a few days later had flown south, probably to the mussel and cockle areas of the Welsh coastline.

By now the insect life above ground seemed to have ceased. Finding a last humble-bee while it hung as if dead on a faded knapweed flower, I touched its furry back and it moved slightly. I brought it into my warm room where it recovered a little and began to stumble across the desk, towards the light. I put it on another flower and its first instinct was to lower its tongue and, with its last dying energy, try to find nectar for the nest. But it was too weak, and as the natural cold enveloped it once more it fell off and clung to the underside of a leaf, waiting patiently for death.

At the end of the month, with the brief spans of most insects over, only a few spiders, beetles and flies crouched beneath leaves, in rock crevices, under tree bark and in the cottage eaves, as if hoping for just one more try at life in the sun. The eggs, larvae or pupae of their myriad species were now ready to overwinter, and of these only the strongest and most securely placed would survive. I picked the last of the blackberries, having culled nine pounds from just four bushes, but I left many for the small birds whose purple droppings showed how much they needed such food.

The long aching winter silences once more lay ahead, the hours of slogging away at the typewriter. Fending off loneliness, I went for twice-weekly treks. I ran or walked across the hills to Shona House for my mail. Coming back on one day, the sky a metallic blue-grey and shot with orange and a crimson underlay from a vanished sun, I felt I was walking in a Celtic or Greek legend. I seemed to see the forge of Vulcan in the sky, leaden, horrible, glowing to the west with an air of latent doom and death, as in the north-east wind I strode, leaping across brown bracken that twisted and snarled at the feet. A huge stag, almost certainly Sebastian, now finished with the rut and pinch-bellied with semi-starvation, leaped with a flash of his light rump patch, stamped on a twig and vanished into the forest gloom. With what ponderous lightness he moved, his clean heavy bulk moving like a dancer. But, as always after a trek, when I reached home I felt better. One is always more creative and alive – spiritually and mentally – after physical exercise. Never more than in a wilderness life is one reminded again and again that the human body and soul are, in this existence at least, really interdependent.

Through the gales and hailstorms of November, finance dwindling, the Canadian book rejected yet again, I kept writing. I decided to aim for a wider audience and sent two stories to

Reader's Digest, plus some notes I'd made when training sparrowhawk Buzz to be free in the wild.

By the end of the month the constant onslaught of the newly swollen burn had carved a new path on the shore, filling the boat parking area with rocks, gravel and banks of debris. It took two days with a crowbar to clear a new landing deck. Shopping trips were only twice a month now. Each time I would come back to another fortnight's total isolation, the outboard burbling away, driving me through an unreal liquid world of almost complete darkness, the inky, oily sea slipping by, the stars above in patches between the hail clouds, the brightest far away to the west above the horizon, and hard granite rocks waiting to hole the boat if one mistake was made. Once I was rewarded when Sammy Seal came really close. It seemed that he had not migrated to the warmer bays further south like some of the other seals, but I had no fish to give him now. By early December even the gulls were fewer, the smaller common gulls had gone south to work for livings round harbours, fields or even city dumps. Just a few curlews, greater black-backs and herring-gulls still cruised widely.

On one supply trip I saw a small flock of 17 white-fronted geese flying high and heading south-east. 'They're away south,' said someone at the store. 'It'll be a hard winter.' But at dawn next day exactly the same sized flock headed over the croft, this time flying *north*. In fact white-fronts spent all the winter in the area, just migrating 10 or 12 miles some days to other grazings. Although many such grazings have been whittled away by development, as there are far fewer geese than a hundred years ago the birds still have a choice of pastures. And certainly that winter proved no worse than the previous two.

With increasing rains, hail and a few days of whirling snow the ground round the cottage became waterlogged and I had to dig a drainage ditch into the burn to take the mountain run-off from my foundations. The garden carrots were now sprouting little white roots, so out they all came, along with much of the cabbage, for a two-day clamping and brining session. Then I dug over the whole garden, turning in the seaweed and compost. As I worked, Little Fat Sergeant, two robins and a blackbird landed in the newly turned soil for the worms and grubs. I noticed that no matter how much rain fell the birds' feathers were always dry. It is not only ducks, gulls and other water birds which have oiled feathers – *all* birds have to some extent showerproof suits. Only when a bird is old or in poor condition do the feathers admit the wet, hastening their deaths in winter. A few days later a hedge-sparrow began coming to the table but the little wrens never did.

One snowy day, my feet cold despite three pairs of socks and hill boots, I was working on the Canadian book, and wondering if I should give up and head south to warmer climes and easier work, when something made me look out of the window. There, creeping along like a little winged mouse through the brown bracken stubs, searching for insects and seeds, was a tiny wren. As it explored the frosty crevices between bracken and grass tufts, it shook little snow showers over itself. Once it caught an old spider's web over its head and cleaned it off with its toes like a cat, moving its foot so fast that it looked a blur. Then on it went again, refusing to give up. At that moment I decided that if the little wren could stick it out so could I. If Robert the Bruce could be inspired by a spider then I could learn a lesson in courage from a wren, and I kept on with my work.

Each Christmas alone I worked out some new little philosophy to help me keep going. And that year I invented a good one. I imagined I had been sentenced to death and that on the last morning the priest came in and said, 'The governor says your sentence has been commuted to life imprisonment'. I realised how grateful I would be in such a situation, despite facing long years cooped up in a jail cell, just to be still alive! Well, I'm alive now, I told myself. And I'm not cooped up. I'm free. So what the hell am I worrying about! And the depression passed.

On Boxing Day afternoon Harry Heron passed like a great cloud by the window and landed in the burn. I hastened for the camera but before I could reach the door he was away again, his broken toe sticking out clearly. I went to see what he had been after but saw only a dipper flying, white rump flashing from its dark brown body, straight up the burn with a faint '*chink*' cry.

I looked towards Shoe Bay – to see a buzzard hovering some fifty yards above the ground, then it dropped straight down like a huge kestrel. I tried to stalk it with the camera but the ground was bitterly cold and wet so I did not crawl, and from nearer than I had judged it to be the big bird flew off. I secured only a brown blur on the film. I found a dead ewe there, ripped about by a fox, buzzards and ravens. That afternoon I built a hide some thirty yards away. Two days later I went into it, but despite a shivering four-hour wait nothing came down. I learned then that working a hide with such a sharp-sighted bird as a buzzard requires at least two people to walk in. Then the bird sees someone walking away, thinks the coast is clear and comes down. She could have been high in the sky and seen me coming from two miles away. I was learning, slowly.

So the third winter on Shona passed. The breakthrough in my wish to write solely about the natural world came in January. *Reader's Digest* sent back the two stories I had submitted but said they were interested in a new one about training the young sparrowhawk to fly and hunt for itself. They warned it was highly unlikely a newcomer would be able to meet their criterion of short but densely focused articles but were prepared to guarantee me a basic fee to try.

No experience in the Highlands had been so complete as the weeks with Buzz. After nearly seven years of wilderness living I was coming up to forty-five. This would have to be the best thing I had ever written, or maybe I had no chance. For days I worked away at it, taking short respites in hikes, jotting down little phrases that came as I walked, my mind filled with the hawk's piercing eyes, his indomitable will, trying to encapsulate the little innuendos I wanted to make comparing hawk and man and all the wild creatures with whom we share the web of life, the chains of time and the splendours of our earth. Yet the words constantly escaped. Writing, cutting, discarding, rewriting, I had it finished after twelve long working days and posted it off.

The wilderness had helped me during a critical period of my life and now I wanted to communicate its great value *for* man in several books. I felt sure that the way to achieve this – as in the Buzz story – was to write from deep personal experience and total involvement. It was this urge that prompted me to intensify the efforts of the last two years to obtain my own wild place to study, in a far larger undisturbed area. I loved Shona, but while the Vanes would let me stay on indefinitely I had no real security of tenure. Above all I wanted a place with trees, for it is woodland combined with open glen and hill that gives the best habitat for wildlife. And I needed a large, wild, uninhabited area behind it through which I could trek the year round with camera and notebook.

After a number of false starts, the memory of the tiny cottage set in its own little woods nine miles up the freshwater loch returned to my mind. Eventually I heard from the landowner that he would not sell the cottage freehold, though he might grant me a lease if the leasehold law changes could be overcome. When I told this news to Allan MacColl at the store one early February day, he said, 'Why don't you go up there when it's fine in the spring and take a good look at it. Ye'll no find a wilder place in all the Highlands.'

Then the thought struck me, I will go and see it *now*, in winter, when it's at its worst. I'll camp out there in the cold, pay the price, get the feel of the place. That evening I visited the estate keeper who said I could use the deerstalking boat which had been left in a small inlet in a pine wood some six and a half miles from the cottage. I had just shopped so I had enough food. I had no tent but there were a couple of sheets of plastic in the Land Rover, and I had a mug, a lighter, two pots and a sleeping bag too. What more did I need? I removed the engine from my boat, stowed it below the vehicle, and with hailstones banging on the metal roof slept the night by the pier.

Wraiths of mist covered the gunmetal sea early next morning and dark clouds drifted overhead but I was determined to go. I filled my pack with the pans, plastic sheets, camera, lentils, rice, two hard-boiled eggs, mince, raisins, bread and three onions, then drove up a dirt-track road past a farm and found the wooden stalkers' boat half full of rainwater resting on the sandy bottom of a small burn mouth. I bailed it out with a pan, fitted my engine, loaded in the pack and extra jerrican of fuel and set off up the loch with a strong westerly wind at my back. It was an eerie, ghostly trip in the near dark and I didn't see another human being or another boat in the full six and a half miles.

The huge old pine snags on the island, remnants of the original forest which had been uprooted by gales, and which had gleamed like whitened bones on that sunny early autumn trip with Dave, now looked different – stark and grey reminders of a primeval age, dark peat showing between their blackened roots. A lone merganser fluttered from the rough water before me into the wind, circled over the boat with odd harsh croaks, then sped downwind, its white wing patches twinkling before it vanished into the misty gloom ahead. The force of the wind increased and I nearly went aground on a long gravelly spit. I fought my way off sideways with an oar before the waves filled the stern and carried on, rolling with the troughs past a long wood of leafless oaks and birches that clothed the steep slopes below the grey and tawny hills. The further I went the darker it became, so that it was sometimes hard to make out where the dark grey of the loch waters met the threatening dank sky. Once I had the feeling that while I would reach my destination I might not get back. My gloved hands were so cold that I had to grip them in the crook behind my knees for warmth.

The loch level was high and as I neared the shore below the cottage I saw the waves falling on to a green grassy bank. I swung the prow into the waves so that it would divide rather than ship the rough water, leaped out and hauled the stern as far as I could up the bank. Even on that dark day the place seemed idyllic. The small L-shaped cottage stood above a forty-yard area of land covered with brown bracken and rushes that lay between two small woods. As I walked round them, counting 21 different kinds of trees and 84 windfalls, ideal for building and for firewood, I found an unusual situation. The four-acre west wood was mostly coniferous, with silver and Douglas firs, spruce and several kinds of pine, while the larger east wood which flanked the main burn and a spectacular waterfall contained mainly deciduous trees, with feathery larch, spreading oaks, ash, birch, holly, rowan and tall, stately beech trees predominating. Here were two separate woodland environments. Clumps of nut-bearing hazel bushes abounded everywhere. I was just thinking how ideal the whole area seemed to be for wildlife when I saw a red squirrel scampering over a ruined wall. The cottage itself, its windows and roof intact, was protected from both the east and west by the woods, from the north by high mountains whose foothills began a few yards behind a small dilapidated woodshed, and was screened from the loch by a fringe of ash, hazel and alder trees which ran along the shore.

There and then I went for a six-mile trek, first to a large river, then up a steep killer of a hill to nearly 1,800 feet, so sheer in places I had to hold on to jutting rocks or small trees, and returned over the tussock-filled plateau along the tops down again to the cottage. I found wildcat tracks in an open patch of sandy ooze by the river. A pair of buzzards spiralled above in the dark sky, the larger female higher and scanning the long view for carrion while the male floated along the rockfaces like a shadow. A single red grouse shot from near my feet and whirred over a ridge, and a kestrel beat past way below, a flickering copper light against the black loch. Two red deer hinds trotted away in a half circle, halted as if to make sure they had really seen a man, their long ears coming forward tardily as I took a picture, knowing that the light was really too poor.

Before dusk I made camp out of the wind in the east wood, gathering wet brown bracken for bedding below one of the plastic sheets. The ideal flat spot was under a huge oak tree but first I eyed it carefully for dead branches above. I jumped up to pull one spike off as I had no wish to be skewered during the night. Then I staked the other plastic sheet down with twigs over three bent hazel wands to make a low tent.

As I wandered through the quiet woods again, the waterfall spouting its white and tawny waters over three forks, the burn tinkling and gurgling through rocky pools so that it was as if I could hear friendly voices amid the music, the wind sighing through the branches above, there seemed an aura of magic about the place. New and old images were coming together strangely in my mind, the great conifers and deep gorge above the falls reminded me of Canada, the hazel bushes fused into the Sussex woodlands of my boyhood, but those images faded and my Highland blood stirred as I realised a feeling of love for this wild lonely beautiful place had already put down roots into my heart.

Was this the end of the long search? I do not know if I spoke aloud or just thought the words. As the old Scots-Indian Tihoni had taught me in Canada, I walked to the trees, touching each one, the oaks, the rowans which are the guardians of lonely homes in Gaelic lore, the beeches and larches. Like the Indians of North America the real old Highlanders had a mystic feeling and reverence for their land, reflected in the poetic Gaelic names for the burns, waterfalls, bays and hills. With Highland blood himself, Tihoni had taught me some of the ancient arts of tracking animals, had told me how the Indians invested every mountain, fall and river with its own deity, good or evil, linked to the great god of all, the Great Spirit. I recalled how his words had made more sense to me than the concept of a single impersonal God, or religions that dismissed all animal life, or the meaningless litanies, chants and catechisms of the churches. I asked the trees, the whole wild place and its spirits of the past, if they wanted me here. It was eerie, misty, and strange; wonderful feelings were coming. My heart was pounding as if I were speaking first words to an admired woman and I think I said, 'If you do want me here, help me'.

It was a hard task to get a fire going. I broke off dry wood from dead branches below the canopy of a huge spruce and packed the twigs with half-dry bracken and eventually got it going between a square of small rocks. I cooked a supper of broth mix, onions and mince, using an old piece of fencing wire for a grille. It began to drizzle. I found it hard to sleep at first, what with my excitement and the noise made by great splats of water falling from the oak branches above on to the plastic that kept me dry. Once I heard a strange thumping and snuffling. The treads sounded heavy and I thought it must be a deer. As I poked my head out I was just in time to see a large boar badger veer off and trundle away through the gloom.

I hauled myself out on my elbows at dawn and was scared by a loud hoffing bark. With my scent obscured by the plastic tent, a small herd of red deer had been grazing nearby and the leading hind had given the alarm. As they cantered away up into the hills a tribe of bullfinches came along, peeping mournfully as they investigated a few old fruit trees, lost among the hazel growth, for early buds. I ate my usual trekking breakfast of cheese, fruit-cake and cereal and watched a flock of tiny long-tailed tits, filling the damp air with high-pitched '*zee zee zees*' as they flitted like flying musical crochets through the twigs above.

Striking camp and putting everything back into the pack, I hiked through the long woods that lay to the west, finding several old nests of hoodies and buzzards in the bare trees. Then I boated right up to the far end of the loch below the high conical mountains and in the strong following winds until I saw the village at its head. When I turned back the waves hit the boat with great force and the engine began to sputter and cut out, as it still often did in stormy water. For the first time I realised why: water was slapping up through the slightly broken lid and was shorting out the plug wiring. How strange that I should solve that mystery in this new place.

Sitting in the centre of the boat to keep the stern up I turned the engine down to its lowest speed, where the spark seemed strongest, and rowed in that fashion for ten miles until I was near the long gravelly spit of land. Making headway was a real battle in the foaming waters, the troughs shorter and deeper than in the sea because of the narrowness of the loch. But I felt the strange and wonderful surge of total solitude, for the loch and hills were totally devoid of human life, as if there was only myself left in the world. Sweat poured down as I heaved at the oars, mingling with the rain on my brow. One man against nature – get there, get there or die, but all you know is to keep battling. Dusk was falling as I pulled into the lee of some small cliffs on which stood a dark conifer wood, just across the water from the island with the giant snag trees.

I camped for the second night in the arbour of the pines and firs, and was woken from damp fitful sleep by a branch thumping down 25 yards from my inert form. After breakfast I went for another long walk in the bare hills before boating back to the Land Rover in the woods. I paid a visit to the green island with its Highland burial ground. On its summit stood the ruins of a tiny church built to the memory of a local saint, and on the stone slab of the altar, washed by centuries of rain, hail and snows, stood an ancient brass bell, green with age. I held the bell reverently, thought again of the little cottage and its woods and silently asked if my work might prove of value. For if so, then I wanted to live there.

Quivering with the cold, wet and fatigue, I hauled the wooden boat out on to the mainland shore, drove to my own boat at the little sea pier, loaded up and, once more fighting the waves, beat my way back to the croft.

Nothing will ever convince me that those three hard days, with 36 stormy water miles covered by boat, 19 miles on foot in the mountains, and two primitive nights out in that lonely wilderness in winter had no bearing on the events that followed.

Within a week the *Reader's Digest* informed me the sparrowhawk story was fine and that I need not rewrite a single word.

When I renewed negotiations to acquire the cottage and wooded area that I later came to call Wildernesse, everything flowed smoothly. By the end of summer Wildernesse became mine – on a lease of 20 years – before the end of which time, I realised with a shock, I would be an old age pensioner.

It seemed my prayer had been answered.

15

Wildcat and Wilder Seas

Spring came early to Ballindona that year. By 7 March the daffodils, thrashing back and forth in the high winds, had grown the green cases of saffron petals that would soon surround the croft once more with a blaze of yellow bugles. In sheltered nooks along the burn banks their rivals, the primroses, were thrusting out the first wan lemon faces for the blessing of the springtime sun. Missel-thrushes filled the dawn air with ringing song, one cock again fluting defiance at the weather from the little spruce in the cliff above the croft. Little Fat Sergeant was back on the bird table and the pied wagtail pair homed in from the east.

On a walk through Shona Forest in calm and glorious sunshine I saw a red squirrel pursuing his mate up the giant fir trunks and along branches in that curious slow-motion chase which is a prelude to mating for these now uncommon sprites of the northern woods. I tried to photograph them, but the heavy 1,000-mm mirror lens had too fine a focus even for their slow chase, and I began to see just how unsuitable it was for such work. The squirrels were dark brown rather than red, probably descendants of the darker continental varieties that were introduced into Scotland when the red squirrel verged on extinction in the early 1800s after many of the original forests had been destroyed for timber, fuel and to make way for grazing lands.

Three days later, with the first humble-bees burring through the air in search of nest sites and newly hatched bluebottles buzzing about the compost heap, I boated out in brilliant sunshine for supplies. At the store Allan MacColl told me he had seen a huge wildcat early that very morning in the fields above the pier where I left my boat.

It seemed a forlorn hope for few humans have ever photographed a wildcat in the wild, and I only had the cheap 300-mm lens with me. Even so I prowled around below a huge rockslide under some towering cliffs in which a pair of ravens were nesting and there, in a marshy patch among the winter-battered old grasses, I found two unmistakable four-toed tracks of a cat, too circular for a fox and too large for a wandering domestic cat. Through the grasses there was a flattened track, too narrow to be a badger's trail and showing no slots from deer hoofs. I went on all fours to play bloodhound but I could discern no taint of fox.

I hid at the edge of a small wood downwind of the area. After a fruitless and cold half-hour my attention was distracted by the ravens. One had flown out when I reached the cliff, but now that I was out of sight it came back with food in its beak, and through my field glass it looked like a chunk of deer meat with some grey hairs still on it. The great black bird hopped on to the nest, a blue-black bill emerged from its depths (certainly the female on her eggs, for ravens are Britain's earliest nesters) and took it from her mate. Instead of flying away again, he hopped ponderously on to a nearby stunted rowan growing at right angles to the cliff. Sidling along it in an oddly undignified way, he started bowing and calling '*krok*' to her, as if still trying to court her, his throat feathers so far distended that he looked as if he were swallowing a large plum.

The raven was too far away for a photo, and it was too dark anyway. His little courting show seemed to have no effect on the female for she stayed deep in the well of the thick stick nest on the ledge, and soon he flew away, beating towards Ardnamurchan peninsula.

Only then did I see the animal in front of me, a mere 30 yards away. It looked almost as big as a roe deer, its legs hidden between the deep grassy tussocks. A heavy, thick-bodied creature with a dark tawny-grey coat, surely too large for any cat, was slipping along through some patches of heather. It paused, lifted its head and then I saw the short muzzle, the broadly spaced pointed ears. It *was* a wildcat, a huge tom, I judged, and it kept low as it headed for some small willow bushes growing in a bog, beyond which were open fields where rabbits had their burrows. It was hunting, and now and again it paused to sniff the air, as if stalking as much by scent as by sight.

Despite the gathering dusk, I tried to take a photo, but the wildcat heard the shutter's click, turned, too briefly for me to take another, and slipped away like a ghost. I stayed unmoving for half an hour in the increasing cold, knowing that while I was fully camouflaged in bush jacket and hat any movement would give away my exact position. I didn't see it again. I noted where it had come from, sure it was using the slight trail from the rock slide, but I refrained from going over to look for its lair as I didn't want to leave my human scent there.

Overnight I camped in the Land Rover by the pier, hoping for a shot of it at dawn in the open area – *if* it used that trail again.

Forcing myself out of the bag in the early twilight, I slid back one of the steamed-up windows to look at the weather. A large otter, its sleek coat dripping water, was carrying a youngster in its mouth across the road just a few yards to the side of the vehicle. But as soon as she heard the sound she dived into the undergrowth. Then I heard a loud plaintive squeak, like the call of a hedgesparrow.

I thought she might have been calling another youngster over but nothing moved.

Without stopping for breakfast, I stole through the woods east of the little road and hid myself overlooking the open boggy patch, ready for a long cold wait. As the light improved I heard a cock chaffinch give its tripping song. Then a pair of tawny owls began calling to each other in the woods behind me. I had never heard owls hooting and 'kwicking' in the morning before.

I had waited about forty freezing minutes before I caught a brief glimpse of the wildcat. It appeared between two large tussocks west of the single-track road, turned south and vanished. I hadn't moved. Nor did I try to take a picture for it was still too dark and there was no point in scaring it. I waited another five minutes, ten, half an hour. Suddenly I saw it again, coming back!

Quivering with excitement and cold, I watched, fascinated, as it stalked along, its head low but moving up and down slightly with every second stride. Occasionally it stopped to peer about and sniff the air, with one paw upraised. It was indeed a huge animal and my heart pounded. Slowly I checked the meter switch. At a thirtieth of a second the needle barely quivered. Treading carefully and yet somehow ponderously, the wildcat came slowly nearer, raced across the little road in a crouching run, as a soldier does when afraid of air attack, then sneaked along through the grass and heather again.

As it reached the open area before me I tried a shot. Even at that distance it heard the click. I cursed silently. But it moved on, mounted a big slab of rock, then stopped and turned round. Perfect! *Click*. It was glaring straight towards me with magnificent green-gold eyes, its huge thick striped tail resting on the ground. Risk a 60th. *Click*. Then it disappeared over a grassy ridge and into the rocky cairns of the slide.

It was the kind of miraculous luck that only comes to a beginner. Later, I used the photo to measure the wildcat's image over the ground – that big tom must have been nearly four feet long.

Knowing it would be useless to stalk such a wary keen-sensed creature, I put the camera and lens into their plastic bags in my pack and boated back to the croft with yesterday's supplies. As I headed north through the narrow channel at medium tide, ready to swing west after avoiding the big sandbanks, I saw what looked like a floating tree in the sea. When I drew closer, I saw that it was a red deer stag swimming, and I recalled the one I had seen in my first November on the island, and how I had wished then for a camera. Well, I had a camera now but it was in the pack! As the stag flailed away from the boat, breath blasting from its nostrils in snorts, I grabbed the pack. But by the time I had undone it, hauled the lens and camera from their bags, fumbled to find the screw thread properly, whipped off the lens hood, set the exposure and shutter speed, made sure the boat was not heading aground and raised the camera and focused it, the stag had gained the land. Shaking itself vigorously, its barrel body and coat colour reminding me of a lion's, it set off at a gallop for the forest below the great Dorlin cliffs. I got a picture all right – of a brown dot with spikes, barely discernible from the gloom of the woods.

Thirteen has never been an unlucky number for me, but on 13 March I was reminded yet again of what an amateur wildlife photographer I really was. I had seen the brilliant black, white and orange oystercatchers in their odd beak-low, loudly pleeping, courtship dashes along the sand and had decided to take pictures of them. I put up a makeshift hide of hazel hoops and fish netting with herbage stuffed in the sides among the deep bracken clumps and after a two-hour wait had focused the bulky 1,000-mm lens on a pair actually in the act of mating.

Suddenly a whoosh of wings and a loud raucous '*Kaah kaah*' right over my head made me jump just as I pressed the button. And away flew the oystercatchers. A hooded crow, almost surely Charlie, had perceived my inert human form through the inadequate herbage on the hide's roof and with the devilish sense of humour some crows seem to have it had dived down and given the loud alarm calls no wild creature ignores. Furious, I leaped to my feet and clapped my cupped hands loudly. The crow turned so fast in terror, thinking it was maybe a gunshot, he almost defeathered himself. Right then I could happily have put a charge of shot into him myself.

By this time three pairs of chaffinches were visiting the bird table, squabbling with robins and each other. Sergeant seemed no longer the dominant cock bird and was having to back down to a new fat cock, perhaps one of his own sons. As usual, with the new grass growing, red deer hinds and last year's calves were coming round the croft at dusk and just before dawn to vie with the sheep for grazing. Once again I noticed that the deer did not crop as closely as the sheep which could still find some food from ground the deer passed over. The hinds were hungry and stalking was easier when they were weak from winter. I was able to photograph nine of them outlined against the dawn sky, wreathed in mist, their heads gleaming like soft golden velvet. I also saw a yearling, or rather a nine-monthling, still getting suck from its mother.

On Sunday I hiked to the big house for mail, but seeing no deer on the low forested route in, I headed back over the high tops to see how far the stags were keeping from the hinds on the lower ground. Spotting a bachelor herd of seven from Shona's highest peak, I decided not to stalk them as mist was smirring, and it would also have meant a long, circuitous stalk. When you're moving downwards deer see you easily against the skyline. But as I walked on and the herd took off I almost stumbled on a couple of stags right below me in a small grassy dell between the rocks. These two, in a small zone of silence, had not seen or sensed the running of the others. They took a few paces, then stopped, as if deciding to stand their ground, and stared back insolently, literally posing for portraits. This habit of stopping for a second look at any source of danger has long been the stags' undoing.

As I looked at these beasts, too close to get even one wholly in the picture, I had a slight feeling of unease for they were big, a good 250 to 300 lbs each, with poor eight-point antlers 'going back' with advancing age. Recently I had been told of a man who had been attacked by a big stag in the rutting season fifteen years before. He had been walking along, unarmed, when the stag had launched an attack, driving him under a small wooden culvert. The beast had almost destroyed the culvert with its antlers and powerful blows of its forefeet before the man had managed to slash its throat with his dirk.

I had laughed the story off as a good yarn. I knew that red deer stags have killed humans but extremely rarely, and usually in parks where they have lost their natural fear of man. But as these two huge beasts glared at me I had a queasy moment. I took two pictures anyway and, as usual, they turned and sped away, floating along, as light on their feet as boxers. I went easily on the hill that day and realised I had had no lunch. It was worth remembering you travel better when light in the gut.

Towards the end of March the sheep were coming closer and, with the lamb season starting, were '*baa*'ing noisily about the croft. Until one got used to it each year, it seemed as noisy as being back in the city. At least this barracking racket (each sheep made a different noise and I swear one had a throat of broken brass) was no worse than I heard on the only radio programme my set was capable of receiving, where there seemed a total obsession with pop music, pop stars, showbiz and football, and whose hosts apparently believed Britain ended somewhere just north of Barnet. So during the early hours each morning, before the sheep went higher to graze, I used the noisy time to make more repairs on the croft.

Unfortunately, good intentions are not necessarily rewarded, and after one long boat and truck journey to the inland town for more hardboard, paint and guttering, I almost failed to get home. The engine dunked beneath the waves in the loaded boat and I had to dry it with cloths under a plastic sheet while rain poured down on to the rough sea. One day, while nailing hardboard up on the rear room wall, I heard a high-pitched bleating. A newborn lamb trying to follow its mother had fallen into the burn and was cold, wet and terrified. I rushed out to rescue it. When I got back indoors I saw a very old ewe licking the lamb all over. I thought this odd as the lamb fed from a younger ewe which was clearly its mother. The young ewe seemed to tolerate the older one's attentions to her bairn, though she butted her rather symbolically once. Iain MacLellan told me that the older sheep would be the young ewe's mother, and that 'granny' sheep often supervised the way their daughters brought up their own lambs.

After three years we had come to know each other better, and when I'd helped with the sheep gathering, Iain had finally decided to call me Mike, which I preferred to the deadly polite 'Mr Tomkies'. He told me that if the old ewes aren't sold off each year in the autumn, and they mate with a tup, they often die. If not during the hard winter then after it in March or April, before or during or after giving birth. 'They're weak from the birth, having to make milk, and the poor winter grazing. The new grass growth also upsets the digestive system. We had a poor gathering last year and lost about ten older ewes this winter, some with lambs in them, which means about £50 less from the sheep.' I thought that few men worked harder for a £50 estate gain than the conscientious Highland shepherd.

By this time I thought I knew the way of the seas well. I was wrong, and on 31 March, I experienced the worst boat trip of all my time on Shona. Setting off in west by south-west winds and hail squalls, I reached the pier without mishap. When I returned to the boat, the gale had increased to Force Nine, the boat had dragged its anchors and was banging against the side stones of the pier. In that violent sea there was no way to re-set it so either I had to go or watch the boat being destroyed.

Banging off against the waves, I hugged the south shore of the loch to gain the slight lee of islets and small land spits. After a quarter mile the boat had shipped a lot of water. A foot-long gash had been made in the hull where the pier stones had torn off the fibre glass patching along the starboard chines. This had not shown up when the boat was empty but with my added weight, the fuel tank and shopping, water was now coming in fast. Twice I had to pull into small bays to pump out the water, my frenzied efforts wrenching the little plastic pump from its fastenings. Hit by a squall of hail, unable to see properly, I was forced to dash into the small mainland inlet opposite the croft. I waited for over an hour, with no overhanging trees or shelter anywhere. Suddenly I realised where I was: this was the inlet that the infamous seventeenth-century smuggler, Black Ranald of Eigg, had pulled into when chased by the revenue cutter from the open sea. He had pulled in behind a screen of high rock and immediately lowered his sails. The revenue crew, believing he had gone up the loch past Castle Tioram, had swept past. Once they were far enough away, Ranald had rehoisted masts and sails and scudded back to Ardnamurchan Point. But this was no time for recalling history. Wet through and bitterly cold, I was shivering, and as I could see no tail to the storm I decided to chance rushing across the strip of open water to the island.

Kneeling in the centre of the boat to keep the stern high, and reaching back with one arm to control the engine, I prayed to the simple god of my childhood – just as I had when in a similar position in a small boat in Canada, only then I had been among a pod of killer whales and still believed the legends about them. Now the waves were six feet high, and short, so more dangerous, their tops lashing off in spray before the shrieking wind. The boat bounced up and down and twisted like a cork as I tried to ride them sideways yet still go slightly *with* them. If the engine had been swamped again it would have been the end, but it held out long enough for me to reach the lee of the island just outside my bay. A near mile of sheer hell!

When finally I hauled the boat out on to its wooden runners my knees trembled. I was shaking with cold and shock as I hung my wet clothes to dry in the spare room. It took two hours to repatch the boat in the next dry period, and I had to gird myself up mentally before my next trip out in it.

The big spring tides were now beating the rocks and the sands lower down, throwing the jetsam higher up the beach as they came in, and on 5 April I saw a motley crew of sea birds working the wrack together. Two oystercatchers, a first-year lesser black-backed gull with a speckled brown back, two herring-gulls and a hooded crow I felt was Charlie were all spaced out. Standing on the edge of the incoming sea, they watched as the waves pounded into the loose seaweed, turning it over and over. And each time the waves receded the birds dashed forward to pick up struggling sand eels, rocklings, butterfish or sand-hoppers. Only once was there any friction between them – when an oystercatcher clearly felt Charlie had come too close, for it leaped into the air and dived on the hoodie, making him flutter further away.

One fine evening Harry Heron flew close to the croft from high up the burn and glided down to the beach. Through my glass I watched him for nearly half an hour – not a movement or stab did he make. I felt sure he was able to live on little but fresh air. I was about to turn away when he made a swift dart and pulled his beak out of the sea with a tiny flounder in it. He didn't stab or spear fish with that great beak, he grabbed them between his long mandibles. Surely, if he stabbed any fish it would be a dab, flounder or small plaice. If a heron stabs a fish he can't open his beak and would have to shake it off – when it could still have enough life left to escape – or scrape it off with a foot, though I had never seen that happen. Though, of course, herons will stab fish to kill them once landed.

Next morning, as I was emptying sacks of molehill soil into the garden compost heap of seaweed and lime, Iain arrived with his collie. He looked concerned.

'I think there's a big, mangy old fox here,' he said. 'I've found two lovely big ewe lambs with their faces torn off.' He had one of the lambs in a bag over his shoulder, its head a gory mess. 'Just born, and a few hours later this happens. Such a fox kills badly, wantonly, drinking the blood from the throat to cure its mange. I've tried all day but can't find its den.'

I had seen the shepherd's life in the hard Scottish hills was far removed from the idyllic myth of a man sitting on a mountain, crook over his arm, smoking his pipe, peacefully whiling away the sunlit hours. I felt that if I were a sheep farmer I could forgive a fox or an eagle taking the odd lamb or two, especially when so many sheep die in winter and lambs suffer accidental death

anyway from various causes in these harsh hills. But wanton killing by an individual fox – where many lambs were slain and often
just left lying about uneaten – was a serious matter. Men's liveli-
hoods were at stake.

I could see the genuine worry on Iain's face. I agreed again to
watch over lambs in my area, and said that if he had trouble
getting the killer fox I would help him do so.

16

The Gull and the Fox

It was on a pertinent date, Friday 13 April, that an avuncular and belligerent co-tenant moved into the croft. A sudden knocking startled me as I was painting an indoor wall. At first I thought it was just the usual sheep rubbing its horns against the porch, but breathless on the doorstep was the island's young schoolteacher. Something bumped spasmodically in a rucksack on her back.

'I found it on the rocks at Shoe Bay' she gasped. 'Do you think it will live? I heard you liked animals. Can you do anything for it?'

She opened the rucksack on the floor and out stumbled the largest and most indignant herring-gull I had ever seen. Hampered by a dangling right wing, it struggled to its feet, glared, then stalked into a corner and registered its disapproval by relieving itself, whitely and massively, on the wooden floor.

Ominous clouds were gathering to the west as the teacher left for home on the other side of the island and I took a look at the gull. His right wing was so badly twisted that the long flight feathers, covered with dirt, tripped him up when he walked. Yet his huge yellow eyes glared at me with jaundiced disdain, like an aging uncle who had been forced at last into an old folks' home. It had seen a few places, this old bird, and weathered many a stormy sky. It looked at me without affection, hope or even fear, just belligerent suspicion.

I spread an old sheet over my desk so that I could examine him but as I bent to pick him up he plap-plapped over the floor on his webbed feet and screeched an ear-splitting '*Hiaow, hiaow*!' It sounded so much like Harold I almost called him that, until I remembered we already had a Harry about the place. I don't know why, but herring-gulls always look like Berts to me, but I'd had a 'Bert' in Canada and had named the crippled gull who frequented the sea shore here 'Bert' too. I couldn't have a third Bert. Well, Gilbert would do!

As I turned him upside down he pecked one of my fingers, hard. I pulled my hand away, forgetting that the edges of a gull's beak, below the red patch at which youngsters pick to make their parents regurgitate food, are like razors. The blood flowed.

Quickly I covered his head with a cloth to keep him quiet and inspected his wing. Holding both wings out to compare the bad one's joint with the good, I was surprised to find the flight pinions overlapped the edges of my desk. Gilbert had a wingspread of 4 feet 8 inches. He was a magnificent specimen, which made me determined to nurse him back to health and let him go.

No bones were broken but there was a slight dislocation, probably due to strained ligaments. The last bad storms had been on 5 April and the injury had possibly occurred then. But how had the gull escaped the mangy, hungry fox that had torn the faces off at least two lambs? Highland foxes often roam the seashore just before dawn to pick up any wounded, sick or young sea bird or dead fish they can find. Gilbert was old and could probably match the fox in cunning by hiding at night.

I settled for the old classic remedy of binding up the wing, immobilizing it, and giving the ligaments time to mend. Taking the one and only bandage from my medical kit, I safety-pinned the wing close to his body, ensuring the pinions were above the tail. Gulls have smooth streamlined bodies and binding a wing is difficult, for the legs must also be left free. Not until I put a coil round his breast too could I make the bandage stay on. As soon as I let him go he tried to walk off, but he kept falling over. It took him two days to learn to balance with one wing tied up. I let him have the spare room. For a roost place I tipped a large box on its side, filled it with grass and bracken, and left a bowl of water nearby.

Gilbert ate just about anything – bread, milk, egg and meat scraps on the first day – then I put him back on to his natural sea diet. He ate like a wolf, and his digestive system was extraordinarily efficient, able to process anything from stale bread to small bones. I had to spread newspapers everywhere.

During the first night I was woken by a loud clonking. I hurried in and found Gilbert on his back. With one wing out of commission he was banging on the floor with his heavy beak as he desperately tried to get up again.

Next day, after checking the nesting herons on Deer Island (Harry's old nest had not been rebuilt and I did not see him), I tried to catch some fish for Gilbert. But it was cold, windy and too early in the year for the mackerel to be in. I had no luck, and when I returned to my beach I collected a pail of mussels, whelks and winkles.

I found Gilbert parading round the room, the bandage flowing loose from his neck like the garland of a seaside beauty contest winner. When I put two mussels before him, he nibbled at them and then looked round helplessly. I soon realised what was wrong. Mussels were far too strong and unstable in a loose state to be pecked or prized open. Normally gulls fly high above rocks and drop them from a fair height. Gilbert could not do this now. So I steamed the shells until they opened slightly and he soon avidly swallowed the contents of half a dozen.

After his meal I took him on to the desk to replace the bandage. When he pecked ferociously at my fingers again I had a brainwave. Bending a pipe cleaner into a double coil, I popped it over his beak as a muzzle. That would protect my hands, I thought. As I began winding the bandage Gilbert was in trouble. His throat was distended and he started choking. I removed the pipe cleaner – and he promptly regurgitated his entire lunch over my writing papers.

I had learned another lesson. When scared, gulls always regurgitate their food. This makes them lighter, so that they can escape enemies more quickly. The great skuas know this and often chase gulls to make them regurgitate their catch in mid-air. Then down swoops the skua to scoop up the easy food. Naturally my ministrations scared Gilbert. From then on I never fed him at all until after his sessions in the casualty ward.

With the bandage back in place I stood Gilbert on the desk. To my surprise he shot forward and gulped all his lunch up again. I recalled that gulls also often swallow all the food they can before competitors reach it, then fly off alone, regurgitate the lot and eat it again at leisure.

I took Gilbert out for a walk on the third day. I felt that sitting down in a room all day was doing him little good. He waddled off down the path but kept tripping and falling over. Gamely he hauled himself back on to his feet with his beak. Looking after an injured gull is like looking after a baby, but you can't put nappies on birds. After twenty odd years as a newspaperman I now had to buy them to mop up after a gull.

He showed no gratitude, of course, and at times I wondered if my care was worthwhile. Herring-gulls are the commonest gulls in Europe, so successful a species that they have been known almost to double their populations in ten years. They have learned not only how to live with man but to *use* him. They have become as much at home on inland rubbish tips, reservoirs, behind the farmer's plough or on airfields – where they are a hazard to jet aircraft engines – as they are on lonely sea cliffs. They are almost becoming pests.

I would look into those extraordinary yellow eyes, like huge glass beads they were, amazingly mobile. He could squint at the tip of his beak, look forward with the stereoscopic sight of a falcon, or sideways with just one eye at a time like most other birds. I didn't want to see that light in them fade and die.

It was not until the end of the first week that I really began to *like* Gilbert. By then the primaries of his injured wing were messy and tattered. Reluctantly I had to clip them just a few inches, after which I took him to the burn to see if he felt fit enough to bathe.

He became totally transformed as he paddled off into the deep pool like a river steamer, joy suffusing his being. He plunged his head in and out with great abandon, throwing water over his back and shuffling his wings. Twirling about like a rubber duck in a bath, he dipped and shook his beak, hunched his shoulders and wiggled his tail, all the while performing a quaint little jig with his kicking feet. He looked so comical I laughed. He wiped his dirty face feathers against his wings, scratched his head with his webbed feet and, with wet feathers, climbed out and up on to the bank.

There he tried pathetically to flap the water from his wings but kept overbalancing because of the unequal weight and the fact that the right wing had almost no movement in it. But heavens, he was trying hard. He preened himself, stroking the water from his breast feathers and shaking it off his beak. In two hours he was completely dry. He looked a different bird, sparkling white, like some advert for soap powder, and his bristly tail was blanched and stiff. From the back he looked as wide as a duck, a small paddle-boat.

If he could make such an effort, ill though he still was, it behove me to do likewise. I tidied up his room, burnt all the dirty papers, scrubbed the floor with disinfectant, gave him fresh bracken and spread out new papers. Then I boiled an egg as a special treat for him.

Gilbert I fear, was becoming an opportunist. Like some overpaid unionized worker he'd now had a taste of the good life and wanted more. And if he could obtain more without an increase in effort, it troubled his conscience not at all. He ate up all the yolk, looked at the egg white with distaste and turned his back.

Next day I left him in his room while I went to meet the factor of the estate that would lease me Wildernesse. We boated there, walked the boundaries of the land I wanted, finally settled on a few acres with access to other woods, and it was dark when I set out across the sea again for Shona.

Entering the croft I switched on a torch. No Gilbert. I looked everywhere. How could he possibly have escaped? Then I heard a faint movement behind some cardboard boxes. There, upside down and wedged tightly between them, lay Gilbert. He must have climbed up, slipped and, with one wing immobilized, had been unable to climb back out. When I put him on his feet, he was dizzy and stumbled about in half circles, but he soon recovered and gulped down a few mussels before squatting down to sleep. Clearly a human room was not the best place for him. If I wanted to return him to the wild the closer I kept him to the wild state the better.

In the morning I took him down to a rock on the shore. As they saw me carrying him down other gulls swooped overhead with the odd '*qwuck qwuck*' duck-like sounds they utter when slightly alarmed or curious. I feared they would attack Gilbert for, like many birds, gulls will often kill their sick or injured brethren – part of the harsh wilderness code to eliminate unfit individuals from the race.

While I watched him through binoculars from my window, only one gull landed near him, perhaps curious at the bandage, as others flew above. None made any attack on Gilbert – maybe because of his great size. Encouraged, I went up the loch on a supply trip, and when I returned, there he was, still on the rock. He had made no attempt to find food. Somehow he had tugged his bandage off again and it was merrily floating away on the outgoing tide. As it was the only one I had I shot out again in the boat and retrieved it. Gilbert's wing, which had always stuck out to the side, like a man carrying a plank, seemed much higher now. The exercises in the burn had done him good.

Later I caught him tugging the hated bandage off again. 'All right, Gilbert,' I said as he padded with protesting qwucks round the room. 'If that's the way you want it, no more bandage.' It was the first time I had actually spoken to him and it had an oddly calming effect, so I kept talking. Gilbert looked up quizzically, his head turning from side to side, like a dog's.

After his bath in the burn next day he climbed on to the bank and flapped to dry himself. I saw with delight that the bad wing now had the right kind of movement in it. While he preened in the sun I left him to do some writing. When I went out a couple of hours later he had gone. I walked round the croft in widening circles but there was no sign of him.

I used the post-dawn search for Gilbert as a photo safari with the hefty 1,000-mm lens, trying to picture the seals on the rocks and a beautiful little roebuck on the edge of Shona forest. Near the shore I found a dead lamb with the fox's bloody marks on its throat, despite a recent visit by the local foxhounds. Clearly the fox had not been scared off the island either and was still killing more lambs than it needed to eat or for its cubs. I just hoped white-plumaged Gilbert really did have sense to hide himself at night.

Searching again after lunch among the thickets of birch, alder and willows along the burns, it seemed that every outcrop of white quartz and lichen on the huge boulders amid the heather looked like a gull. Finally, almost a mile away, as I returned along the rocky shore line from the west, I found him. He cried out '*hiaow hiaow*' in alarm at my approach and led me a hard chase, leaping from rock to rock and spreading out his bad wing with the other to lessen his falls. At least he was in good shape. Eventually I caught him when he tripped on seaweed within inches of the sea and escape by paddling. I brought him home for a whelk and mussel supper.

To keep him outdoors yet within checking distance, I made some thick leather jesses for his legs and put him out on a long line, as I had with the sparrowhawk Buzz. But Gilbert was no hawk. He could not learn to ignore what he could not change, and tugged persistently at the jesses. During his usual spectacular bath he twisted the line around his legs half a dozen times. When he saw me coming he tried to run but, hampered by the lines, ended up performing an undignified nose dive into a patch of mud.

He shook his beak furiously – gulls hate getting dirty – and when I picked him up he pecked so viciously that I was forced to fetch gloves and his pipe cleaner muzzle. Finally I snipped off the offending line and jesses and Gilbert waddled disgustedly down to the beach. He found a rock crevice filled with moss and grasses and settled down for the night. Each time I looked through the field glass until dusk his white head was visible, peeping out just above the rock, like a white golf ball, as he kept a good eye on me too. I felt sure that if the fox left delectable young lambs lying about it would hardly bother with a tough old gull like Gilbert, who could inflict some damaging pecks before he succumbed. I decided to leave him alone for two days.

On the morning of 24 April I saw him standing at the edge of the incoming tide. As each wave rolled in it nudged the loose seaweed a little further up the shore and Gilbert darted forward to grab sand-hoppers and small sea creatures. Each night I looked for him to make sure he had a shellfish supper. It became clear that he knew he had to hide for I always found him deep in clumps of new bracken, thick rushes or beds of yellow flags.

The sun shone so strongly through the window next morning that I took down the plastic tent around my desk and stowed it away. Outside Gilbert had disappeared but I didn't look for him in such good weather. Instead I sieved garden earth and was busy planting vegetable seeds when Iain appeared. He had his shotgun and two little bedraggled terriers, a cairn and a Highland, which he had borrowed from a mainland friend. He said the big mangy fox was still killing; they had lost 23 lambs to it this year and he was now going after it himself. The cairn terrier was on a lead, for it wasn't safe with lambs about. I looked at the other terrier, such a tiny scrap of a canine it seemed, with watery doe eyes and a woebegone expression. I suggested it could hardly be a match for a rat, never mind a fox.

'Aye, it is,' Iain protested. 'It's accounted for a score of foxes already.'

He lifted the terrier's lip to expose canine teeth almost an inch long! This tiny sweet-looking creature had in fact ripped Morag's cairn bitch when Iain had brought the duo over in the boat the previous day, making a deep, two-inch-long gash in its back from which it took many days to recover. I felt there was reproach in Iain's eyes at this semi-Sassenach's apparent pacifist attitude to foxes. I fought a losing battle with myself for I knew two guns were better than one when after foxes at the den.

'Have you got a spare gun?'

'Aye.'

'Right. I'll come with you tomorrow.'

To my surprise, far from showing gratitude, Iain seemed worried about my handling of a gun, that I might shoot one of the terriers in mistake for the fox! It was easily done, he explained, the cairn being of a similar colour, the fox bolting suddenly, the heat of the moment . . . One of his own co-workers had shot a fine terrier only a few years back. As Iain talked, memories of my younger days when I had lived with guns and shooting came flooding back, the years with a fine gamekeeper and my father, both superb shots, on a 4,000-acre estate in Sussex, years of ferrets and nets, long netting in the fields, pheasant drives, rough shoots, deer culling . . . the times I had hit running rabbits with rifle bullets, and for a bet had once taken two partridges out of a covey, both through the head, with a 22 bullet. I recalled the weeks of marksman's training in the Guards, the Canadian wilderness years, trekking through bear country when one had to shoot a deer or two, purely for sustenance. But I said little of this to Iain for I was no longer interested in shooting animals and, after such a long interval, was by no means sure my performance next day would match any past skills I may once have possessed. Instead I assured him I could handle a gun. I also pointed out that it was correct always to break a shotgun when crossing any obstacle, that one should never put one's thumbs over the barrels, and when one left a shotgun leaning against a cottage wall it was a good idea to remove the cartridges first.

Next day I boated to Shona pier, picked up the spare 12 bore and was told to join Iain on the high hill north of the main forest. Sure enough he was waiting there, right at the top. I hurried up so as not to keep him waiting and as soon as I arrived, puffing, off he went like a rested deer!

It was a great day on the hill as we traversed the island, working the corries, the major rockslide cairns and the dens Iain knew. Usually we stationed ourselves forty yards apart, covering wide arcs each side. Only at one den with many exits on a precipitous 500-foot slope that ran almost sheer down to the shore near Deer Point did the terriers hole up for a while. First there was silence, then the sound of little running feet below, then silence again. Twenty minutes later both terriers emerged. Iain thought the vixen was away but they had probably killed some cubs.

When we paused for lunch on the leeward side of a hill, the spring sun lighting up the greens of new grass against the grey and tawny winter drabness, Iain told interesting yarns of his battles to keep foxes down before the use of both the gin trap and leaving poisoned carcases on the hill had been banned.

'We used strychnine in a dead cormorant or in hens' eggs. It had to be the right amount – too much and the fox would vomit immediately and the poison wouldn't work. For the eggs, I broke off a bracken twig, dipped it into the strychnine and then stirred it briefly into the yolk – just enough. As soon as the fox feels the poison he rushes to water to clear his insides. But the water expands and helps kill him quickly, like rat poison. Now, one winter I put a cormorant's body on a rocky islet just off Deer Point, with two eggs by the side of it.

'That year we had a tup we could not catch in the autumn gathering, and the following April I went back there with Morag, found the tup on the Point and decided to drive him to the old schoolhouse on the north shore so that we could fetch him by boat the next day. I had to get above him and work him down, because if he got in front of me he'd take off over the hill. But he wouldn't keep to the path – he ran down west and on to the cliffs.

Then I could only come at him from below for he was on a tiny ledge. I climbed up, but my only handhold was a clump of heather. I grabbed his horn with my right hand, scared that if he jumped he'd pull me down too. Well, he slipped, and then he was dangling from my hand by the horn. It just showed how starved and thin he was, just a bag of bones, if I could hold him up by one horn. I lowered him slowly to Morag who got hold of his backside, and we got him down the rocks without injury.

'As we herded the tup along Morag suddenly pointed to the islet off shore – and there was a strange sight. The fox had got there, eaten part of the cormorant and had then felt the poison. He had just made it across to a small burn, drunk water and had died right there. His skeleton was on the bank, his big tusks showing white even at that distance.'

Iain told me of a young keeper who tried new ways to catch foxes. 'To save on cost, he made a snare from hay bailing wire and we all laughed at him. One day he came in and dropped a fox in that snare at our feet. The fox had got its head and paws through but was caught round its middle, and in its efforts to escape had torn its insides. The snare looked like the same wire the fence was made of – and the fox was not suspicious of it.'

He paused, a sandwich in one hand as he smiled at another memory. 'The same keeper made a big cube cage from wire netting and fencing stobs. Then he made a hole at grass level in one corner and baited inside with a dead lamb, pieces of venison, fish and a dead bird. He left it several nights, but once he found the fox had been in and eaten some meat, he set a snare over the hole. He was as clever as a fox himself. Instead of first setting a snare and making the fox suspicious, he let it think it could get in and out easily for free meals, *then* set the snare. He caught nine foxes that way.'

I asked Iain if he agreed that folks who did not lock their hens up at night in a properly secure poultry house really deserved all they got?

Iain laughed. 'But what do ye do when a tree falls on your hen house? That happened to my mother's ducks and hens. Until we could repair it we put them into a shed which had housed the dry toilet. The door was kept shut but we made a small hole in the bottom so that the hens and ducks could get in at night. One morning there was chaos. Nine hens and ducks had been killed and the fox had buried the ducks and covered them up with leaves. But the odd thing was he buried them all upside down so nothing showed through but the bottoms of their webbed feet. These pairs of little webbed feet were sticking up everywhere, as if the fox had left them as markers for his return.'

It was back at his house after our tiring and foxless fifteen-mile trek, and after Morag had cooked us a fine supper, that Iain handed me a good dram and told me the oddest fox story of all.

'Did I tell ye about the den we found through a dream? We caught an old vixen in a snare once and she was in milk, her teats full, so we knew she had cubs, but *where* were they? Two nights later I had an odd dream – no-one believes this but it's true. I saw a sand hole I knew of and in the dream the vixen came over the hill, a lamb thrown over her shoulder, and went into the sand hole.

'Next day I went straight there. There were now three holes, and one of them was scraped out new. I put the terrier in the top one and he came out of the second, went back in and came out of the bottom hole. Naturally I thought there was nothing in there. Then I saw something way down in the hole. I cut off a small forked branch for a rake, and the first thing I scraped out was a heron, still with seaweed round its feet from where the fox had caught and dragged it. Then one by one out came six lambs. All had their rich milky entrails taken out for the cubs.

'We found a dead cub buried a few yards from the hole. I scraped away and out came two more, also dead. The funny thing was *all* the cubs had bloody scrape marks round the ribs. One rib was stripped bare on each side, almost all round the body. We couldn't understand it, but after talks with friends we decided what had happened. The cubs, hungry for milk after their mother failed to return, kept wandering outside. And each

time the dog fox had carried them back to the den. He was killing
for them – a heron, lambs and so on. But they were too young to
eat, they needed a drink of milk. And each time they went out
looking for their mother, so he carried them back. Foxes have
long canine teeth and after hours of this his teeth must have worn
a girdle round the cubs' chests for all were bloodied in the same
place. Finally he must have buried the cubs himself.'

As I clambered the two miles home in the cold darkness I felt
a little guilty. With Iain's critical eye upon me, if a fox had made
a bolt I would have shot it. He worked hard to preserve lambs in
these harsh hills, and if he said 23 lambs had been killed by an
unusually wanton fox I could believe it. To remove from future
generations the genetic strain of such a killer made sense. What I
could not believe was that *all* foxes kill in such a manner. In any
case, how would we know we had killed the right fox? I still
wondered if after such disturbances as the hound pack there
might not have been an element of revenge in the killings. At the
time, however, I knew little of such matters from the Highland
shepherd's point of view, and I was not too sorry I had not shot
a fox.

17

To the Rescue

When I had not seen Gilbert the gull for three days I started tramping the rocky cliffs and beaches to look for him. I was over a mile from home when I heard the quivering bleat of a tiny lamb. It was lying alone by some ruined walls and there was not a ewe in sight. As I drew near it staggered to its feet and tottered away from the huge giant it had never seen before. It was pinch-bellied, hump-backed, so it had probably not been fed much since birth, if at all.

I rounded up the nearest ewes and drove them towards the lamb. Bleating pitifully, it staggered towards them, but as it sought the teats of one ewe she swung round and bunted it off its feet. It was apparently not the mother and the other ewes ignored it.

I could do nothing but bottle feed the lamb and, when it regained its strength, try to put it back to its real mother or a foster parent. I slipped the thin, shivering mite into the backpack and walked home. There, who should be walking up the path but Gilbert!

He looked frayed about the head, his eyes sunken, and he was hungry. The sea had been calm for two days despite a mixture of hail, snow and sunshine and had not helped him by turning over the seaweed. I was delighted he had walked all the way up to the croft when things were going badly. As I fed him meat scraps and cheese – which he loved – he kept away from me. He was warier now after his freedom, stretching his neck out for the food but looking at me as if to say, 'Keep your distance. You may feed me because you have been appointed to do so by the Great Sky Spirit but don't assault my dignity by picking me up!'

For three days I struggled to keep the lamb alive, handling it with gloves because if I did succeed in finding its mother she would identify it as her own by scent. Iain came by once and assured me I had done the right thing by taking the lamb.

'It's always a dilemma. You leave it and when you go back the next day it's dead. It probably had a young, inexperienced mother and never got a suck of milk.' He shook his head when he saw how the milk had to be forced into the lamb's throat by squeezing the plastic bottle. 'That's a bad sign. If it took to the teat after tasting the first milk ye'd be all right. But remember, if it survives, it will be like a little dog and once it has your scent it will follow ye everywhere. Ye'll be stuck with it for three months.'

Gilbert stayed near the croft each morning now, as if he was jealous, for he had never done so before. Sometimes he woke me by turning the debris of my compost pile or rooting among my box of old tins. He even acted as an unwitting decoy for other gulls and some mornings I woke to see flashing white pinions outside the window as a small flock joined him at the heap. As soon as I had fed him he sneaked off down to the shore again, sidling along under the herbage, his head low, like a furtive white rat. Once fed, it seemed he wanted to keep out of my sight in case I caught him and took him back into the house. Usually he stumbled over the stones for a drink of sea water, a swim or a refreshing bath at the foot of the burn. At night he would come back up again, find a hidden grassy patch in the lee of a rock so that the wind would not blow up his feathers and freeze him, then settle down to sleep.

I admit I neglected Gilbert a little at this time. The lamb was delicate and hard to keep alive. Sometimes it recovered enough to climb into my waste paper box or into the fireplace and look up the chimney. It sniffed at my shoes like a dog and at all things under the bed. When it scented objects its top lip curled upwards like a black bear cub's. Occasionally it scratched in my seed box of tomatoes and ate the fine earth, as lambs do, or followed me from room to room, its tiny black trotters going '*donka donka donka*' over the wooden floor. Then I would find it lying down, eyes closed and jerking or coughing convulsively. I lit the fire for it, but it seemed I had arrived too late. The cold April nights, combined with lack of mother's milk, had given it no real start in life, and on May 3, it died.

As sadly I buried it I realized I hadn't seen Gilbert for a while. When the other gulls fluttered up from the garden tip around dawn no earthbound gull remained. Had they chased him away? I looked down to the beach and saw the tide was up, so I decided to search for him by boat.

Half a mile away, where the island reached out into the open Atlantic, I found him sitting on a huge domed rock, his feathers fluffed up and his eyes sunken again. When I went to catch him, his plumage sleeked down and he fluttered away, actually flying a few yards before plopping down heavily into the water. Clearly his bad wing was stronger now though he was still handicapped by it being shorter than the other. He did not look well. As I tried to get near him he paddled round the boat with a curious zig-zag movement, as if one webbed foot was pushing harder than the other. He made it to a sloping rock but could not get out of the water. He flapped, trying to scale the rock's wet surface, his left foot hardly working at all.

He pecked hard when I picked him up. When I turned him over I was shocked – one of the toes on the lame foot had been almost severed and was bleeding. The next toe was badly cut and the dew claw further up the leg was missing. He must have been in pain and, almost unable to walk, had found it hard to get food. It seemed the fox may have caught up with him after all, with Gilbert only escaping into the sea literally by the skin of his foot.

That was the end of Gilbert's freedom. He couldn't look after himself now. I clipped on his beak pipe cleaner and carried him home. Now it would take as long again to get him back to health.

As I walked up the path with him tucked under one arm, passing the profusion of primroses lining the burn, I saw that the rowan trees were the first to bud and throw out tiny silvery leaves, followed by birch, alder and hazel. The sap in the oaks and ashes would take longer to stir, and they seemed to be still in winter's grey grip. The first cranefly flew uncertainly by, tangling itself briefly in new bracken heads. A black dor burying beetle, just out from hibernation, burred clumsily by and three humble-bees dosily worked the daffodils and primroses. From afar came the soft call of the first cuckoo. As we entered the croft a swallow came homing in from the south. Twisting and turning as if with joy at being home again, it shot past us and headed north. Gilbert's bright eye followed it all the way. I thought he had an almost wistful look, for he too had once been a master of the sky.

Indoors I put Gilbert on his back, bathed his injured feet with hot water and diluted Dettol and then put antibiotic cream on the wounds. After a treat feed of raw steak and mussels I left him to sit back in his old room. He seemed glad of its security and was far tamer, as if he now realised I was really the only friend he had. There was a softer look in his eye and for the first time he let me stroke his snow-white neck and sleek grey wing feathers, which he still insisted on cleaning though he could now only stand on one foot. It was a marvel to me that, no matter how ill he was, Gilbert was fastidious about his toilet. Perhaps keeping himself spruce was the only morale boost he had left.

Four days later, however, his foot and knee joints had swollen to twice their normal size. In the burn he made great efforts to clean himself but the bad leg stopped him climbing out again. I had to set him on the bank where, hopping on one foot with the swollen leg raised, he still tried to preen while constantly being blown off balance by the breezes.

In my attempt to prevent the poison spreading I fed him luxury foods – egg yolks, steak, whelks, mussels, cheese and even a few prawns I caught in the small baited funnel trap. I also put vitamin C in his water. I didn't want him to lose all movement in the bad leg, so for a time I herded him round outside like a sheep as he 'qwuck qwucked' in protest, but before long he just sat tight and faced me with open beak. One morning, despite the rain which had swollen the burn, Gilbert was so dirty and woebegone through loss of morale and lack of exercise, I found a quiet place in the deepest pool and left him there. I felt the flow of the water would force him to swim, keep him exercised and keep the leg muscles working.

Foolishly I left him alone for too long. Suddenly, as I worked at my desk, something white caught my eye through the window. Gilbert had somehow swum into the mainstream and was being washed over the little falls. I rushed out as he tumbled like a frayed wrap-tangle of old rags. I thought he'd be dashed to pieces, but somehow he recovered his footing twenty yards downstream and was flutter-staggering out, bedraggled and wing-heavy, as I reached him. I had clearly over-estimated his strength for he was now very weak and shivering with shock. Tenderly I carried him to his room and fed him. Too weak now to stand at all, he *still* tried to preen and dry himself while sitting down.

Over a few days of rest the swellings began to shrink, but when I turned him on his back, although he pushed up indignantly with his good leg, the other had lost all power and two of the toes were inert. It was serious, for gangrene could set in. I tried taping up his good foot so that he would have to use the other to move. But he just sat, pecked at my efforts to move him and finally ripped off the tapes with his beak.

I now had to supervise short sessions in the burn, scaring him round in the water to try and get the bad leg working. At night I held bits of his meal away from him so that he was made to come to me for them, but he only raised his wings like an old plane trying to take off and hopped on one leg. There seemed only one hope left – physiotherapy. For the next few days, a quarter-hour at a time, I laid him on his back and exercised and massaged his foot with my hands. Anyone peeping in at the window and seeing me talking to and playing 'footsie' with a gull may well have concluded I was a suitable case for some different kind of medical treatment myself!

To keep his wing muscles toned I walked and trotted through the nearby pastures holding Gilbert's body high in one hand like a model aeroplane. Soon he was flapping away with those regular jerky downward beats that gulls use when effortlessly keeping up with boats and watching for thrown scraps from passengers. He really seemed to enjoy these sessions.

One day Iain came round while I was exercising Gilbert. He watched me awhile, then thrust his hand deep into a bag and handed me a fine two-pound sea trout, for which he said there was not room in the deep freeze at home. As I thanked him he looked round, his blue eyes shining brightly.

'You've made a good life here, Mike,' he said. 'I sometimes wish I had this place out here, just to sit and do some thinking, away from everything.'

I looked round the cottage, seeing it again as if for the first time, noting I still had not properly repaired the porch door.

'I haven't done as much with it as I should have,' I said. 'But I'll fix that door before I leave.'

'Ah,' said Iain, paying me a compliment which, coming from him was one I will never forget. 'You've brought the old place to life again, Mike.'

It was five weeks to the day since coming to the croft that Gilbert's bad leg showed signs of returning to life. To perk up his appetite I had put him in the burn for a bathe. When he had enough, he reached the side but still could not scale the steep bank, so he just clung there with one foot, his outstretched wings propping him up. Cruelly I just left him and dangled his favourite raw steak above. Finally he struggled up again – then I saw it. The bad left leg went down and pushed. Then he was up on the bank and eating the steak. Later, when he preened himself, he put the bad foot down tentatively for balance. The crisis, it seemed, was over.

During the next few days I kept Gilbert to strict play periods chivvying him over the sunlit grass and daisies outside and making him flap and run after long squatting periods. I tempted him with titbits, threw him a few feet into the air so that he had to fly and land on his own, but always rewarding him with food. He rarely pecked now, as if he faintly realised I was only doing it for his own good.

Towards the end of May, the buds finally breaking open on the ash trees lining the burn, it was clear there had been a good surviving lamb crop, despite the earlier predations of the old fox The sheep near me grazed on an oval daily migration, starting from their sleeping spots high in the hills and working down westwards to the beach. By midday they were all round the croft, and at that time of the year I was treated to a real lambs-and-ewes chorus as each had a voice of slightly differing timbre.

One day I was watching Gilbert have a good preen after his dip when a baby lamb with a fine wide face and long black boots went up, sniffing cautiously like a dog, intently curious about this great white bird. Gilbert, scared, jumped up and flapped his wings. The lamb leaped about five feet into the air and came down on the wrong side of my garden fence. It dashed over my new lettuce plants to escape this frightening feathered creature, and as it could have damaged itself against the wire netting I rushed out to rescue it. Off it went, shaking its head and doing a high skittish leap every so often, until it found its mother's reassuring teats. Then it banged her up and down like a boat as she stood there looking at me with happy, silly-looking chewing smiles.

I soon found that most of the lambs liked to come and sniff Gilbert. Sometimes they stood in small groups, heads close together, as if discussing what prank to play on him next. Then up they would come, Gilbert would leap wildly, terrified, and away they would flee, hugely enjoying their little fright. As the lambs liked to frolic in and chew the fine earth from my seed boxes, I put Gilbert on duty beside them like a guard dog. That I had any onions at all that year was largely due to Gilbert's success as a sentry.

The scrub oak trees were at last in leaf on the day I decided to give Gilbert his freedom. I carried him down to the sea and, facing the warm southerly breeze, gently threw him into the air. He spread his wings and flapping harder than gulls normally do because his wing tip was still short, he flew round my head.

The sun shone through his pinions, giving them a beautifully transparent look, and as he passed above me his white head turned to look down. '*Qwuck qwuck qwuck*' he intoned softly almost as if saying thanks. Then away he went out to the far rock of the bay. There he landed in the water and began happily paddling around.

He occasionally came near the croft in the first few days, snapping up food I left for him, and sometimes at dawn I saw him down on the beach, hurling over small piles of seaweed with great gusto as he searched for breakfast. By late August new primaries were growing to replace the few I had been forced to clip so that he could then soar and wheel almost as well as the other gulls. Sometimes he came over my head as if allowing me to take photos of him back in possession of his full flying powers.

He no longer allowed me close enough to pick him up, but as the weeks passed by and I looked out in the mornings and saw he was still using my bay, where most of his battle for a return to health had taken place – well, that seemed reward enough.

18

Talking with Ravens

For the rest of my last spring and summer on Shona I tried to intensify wildlife observations and spent many cold nights out in uncomfortable places. Watching wildlife is difficult enough but successful photography in the real wilds, when major mammals are most active at dusk and before dawn, is harder still. And in these open hills the glens, rockfaces, long ridges and low corries so changed the course of ground winds that they often blew in the opposite direction to the movements of the clouds.

My first target in May was the big fox which I thought was hunting mainly among the rockslides, meadows and a huge half-mile-long corrie between the high hills and seaward bluffs on the west edge of the island. Whether the terriers had killed a cub or two in the steep slide by Deer Point or not, I was sure the vixen had taken any surviving cubs elsewhere and that the foxes would be using other temporary dens in the area now. This idea seemed to be confirmed when Iain appeared by the croft on 21 May and said he had been on a foot hunt with the foxhounds. They had found a warm freshly killed tup lamb above Baramore on the north edge of the island, its innards taken, possibly by the vixen for a cub or two. The hounds had followed scent to various small corries and slides near Deer Point but had not been successful.

I told Iain I would set baits out for the foxes and keep watch overnight. I felt Iain was right when he suggested that baits only really worked in a harsh winter when foxes were starving. I said there was probably too much natural prey about now.

'Aye,' said Iain with feeling. 'Too many defenceless young lambs about now!'

Still, I felt it worth a try. I made a hide of hazel wands, bamboo and camouflage netting stuffed with bracken and grasses and, after boating for supplies and a pile of meat bones and sheep ribs from the butcher, I rowed to Shoe Bay and carried everything over the hills to the big corrie. Choosing an open spot near some large boulders, so that the fox would feel safe, and wearing gloves dipped in spruce-needle juice, I set the meat and bones out, pegging them down with ash twigs and obscuring these with scattered earth and grass tufts. Then downwind and about 90 yards above, I set the hide up between three five-foot-high granite slabs. Sure the fox would not approach the bait on the first night, I went home.

Up before dawn two days later, I trekked back to check. On the way, as I slipped between rocks on the western shore, I saw an otter swimming in the sea, rolling and playing among the fronds of brown kelp that washed to and fro on the ocean swell. Although the light was poor – I could barely see the otter through the dark mirror lens – I took two photos (hopeless as it turned out) and carried on and into the hide above the baits.

I watched as the light grew stronger but no fox appeared. It seemed the baits were now in different positions. When I went down in full light I was sure the fox had been there. The bones had been torn from their fastenings, most of the meat eaten and when I bent low I could detect a musky scent. I reset the red ribcage, added a juicy new bone and left.

Returning well before dusk, I found the bones had again been ripped away. Ravens were pecking at the empty ribcage, and the smaller bone had disappeared completely. So had the fox come to the bait in daylight? I resolved to wait out all night, and after re-setting new bones (now slightly off as I had no fridge), I retired to the hide. Hours went by as dusk turned to almost total blackness. Only the miserable discomfort of my sitting position, the cold of the night, increased by the cooling sweat in my clothes,

kept me awake. Both camera and my field glass were tied in position to the hide, and were supported by one of the rocks. It is an
eerie business, sitting over two miles from the nearest human
habitation, shrouded by darkness in a gloomy glen, the huge
hunched shapes of the hills brooding all round you in the land of
the fox, eagle, wildcat and stag. The eyes play tricks in the dark,
images form and pass across the retina of your mind and you
have constantly to move focus, blink and look away and back to
see things at all. The whitened bones were all that could be seen
through the field glass, hardly at all through the camera, and at
times appeared to be moving on their own. Many times I imagined I could see the form of a fox but when the pre-dawn twilight
came it was clear the bones were exactly where they had been and
were untouched.

Then the ravens came in. First a loud '*krok krok*' sounded
above me, then down swept the first huge black bird. It landed
with a sinister bat-like folding of its long broad wings, kept
deadly still for a few seconds, head and powerful wedge-shaped
beak held upwards, then stalked sideways round the bones. Its
first pecks were attended by brief jumps backwards and scared
looks around, as if it anticipated a trap. Presently others came
down and all began tugging at the meat on the bones.

It was the first time I had been really close to so many ravens
and I was amazed at the different sounds they made. I tried to
divine the meaning of some of them during the next two hours
and compared them with other raven noises I'd heard over the
years. The loud '*prruk prruk*' seemed to be a light warning call in
flight, a loud '*roawr*' was a serious warning to others, often made
by a flying raven or one perched on a rock when its mates are
feeding below. There was an oddly metallic '*poop poop*' sound,
rather like a loud cuckoo, seemingly made to inform a mate that
it was in the same vicinity, an amusing '*gloop*' made when
performing aerobatics and flying along on its back, and an even
more hilarious metallic '*boing*' when a raven was turning on its
side in flight, going into momentary free-fall, and uttered just as

it righted itself again. As they fed now they hustled each other out of the way and made a variety of soft sounds – '*glogs*', '*whows*', '*crucks*' and '*pops*', but what each of these meant I had little idea. That ravens talk to each other and have a large vocabulary was obvious. But why should such a variety of sounds have evolved in a bird that is not, after all, as gregarious or a social nester as the rook or the jackdaw?

The sun was over the hill and the ravens had all gone by the time I roused my stiff body and went home for some brief sleep.

As I had much writing to complete, I decided to make only one more attempt at the fox and was back at the big corrie with the last of the meat bones at 8 p.m. This time the old ones had not been pulled about but had been heavily pecked by the ravens so that there was little meat left on them. Just in case any fox was watching and to obscure my scent, I kept some yards away and threw the new bones by the old. Then I walked over the hill towards the shore before doubling back upwards and creeping into the hide, keeping low between the rocks by following the course of a freshet.

It was an even colder night after a gorgeously hot day, the first of summer, and the sky was clear. At least the stars shed a bit more light on the scene though it was still extremely dark. At about 2 a.m. I roused from a brief doze to look through the eyepiece of the camera. I distinctly saw something move across the grey image of the bones. I stretched up to the field glass which had better light-gathering power and blinked with excitement. The fox was there! At least I thought it was, for there was a dark shadow, doglike but slimmer and longer-legged, moving by the bones. It moved slowly, as if floating, and I had constantly to shift my gaze slightly to make its shape out against the dark background. Certainly it was a fox, a big one, and I could now see its huge brush, almost as long as itself, thin at the base and swelling out like a black banana before thinning down to a point at the tip. As I watched, fascinated, hoping it would not hear the banging of my heart, it seemed to grip the last big bone with its mouth and sink down until its belly was nearly on the grass. Then it tugged with great jumping jerks, its forefeet braced far forward. The bone must have tangled up with some herbage or the ash stakes. Occasionally it stopped pulling and its front changed shape as if it were moving its head, watching and listening.

Then back it went to the attack, pulling with strong jerks. I slid down to look through the camera but could not make it out against the general gloom through the lens. Back to the field glass. Within a minute or two it had worked the bone loose and, knowing it would run off with it, I took a chance in light that was far too poor, and pushed the camera button hopefully at an eighth of a second.

Instantly the fox stopped, froze dead still, and suddenly was no longer there. Even at 90 yards upwind, it had heard the metallic click of the shutter and was away. That was that. I knew it wouldn't come back that morning, probably never. But I kept waiting. Once again as light came so did the ravens. One by one they swooped down like monstrous black bats and again the corrie was filled with their strange echoing voices. This time I was treated to an odd little spectacle. Two of the birds started walking alongside each other, their throat and head feathers all fluffed out. They turned and bowed to each other like contestants in a Judo contest, one bird gave a peculiar '*chrraw*' croak deep in its throat, its head feathers flattened down except at the sides where two distinct crests like little horns appeared then it jumped at the other with its claws extended. The second bird immediately flew to the far side of the group.

As I neared Shoe Bay on my way home I again saw an otter. Two days later, when I took the fox hide down, I set it up again 40 feet from where an otter had left fish scales and bones near a freshwater pool. I spent two more early-morning vigils there but not another otter did I see. It was a long time before I learned one should never use hides for mammals like foxes, otters, badgers or wildcats. Hides are mandatory for birds but are useless for animals with not only good sight but a keen sense of smell and higher reasoning powers. With mammals one is far better off arriving surreptitiously downwind in camouflaged clothing and just waiting amid trees, bushes, bracken, long grass or between rocks.

There were minor compensations during this tiring period. Once I came back to find a merganser drake and duck preening themselves in brilliant sunlight on a rock just above the sea, and took some pictures. Another time I saw a mother eider fly up from heather near the shore. She had seen me coming from a fair distance and had covered her seven dull-green eggs with her dark-brown down. I set up the hide but spent a fruitless afternoon. With her eggs covered in this way on such a hot day she had no need to return for several hours. I was about to give up when a hooded crow flew to a rock near the nest. This will be interesting, I thought, a picture of a crow actually taking an eider's egg. Suddenly, with a great wooshing noise the eider mother swept in, landed with a loud crash in the heather by the rock, scaring the crow out of its wits, for it leaped into the air and flew off almost as fast as a hawk, squawking with fear. Clearly it hadn't seen the eider coming and it wasn't going to argue the toss! None of this did I photograph. I was focused on the nest alone.

I also found that if I kept a sharp eye on the ground there could be extra bonuses to every foot trek. On a sultry walk to the east I found a huge blue-black beetle, the like of which I had never seen before. It had tiny short elytra, or wingcases, but no wings at all, and as I picked it up it released pools of clear orange-coloured oil from its joints. I popped it into a specimen box and, when consulting my insect book at home, found it was a rare oil beetle, which could measure anything from 13 to 32 mm in length. This one, then, was a giant for it measured a fraction over two inches! Its life history was more fascinating. The female lays batches of tiny eggs in holes in the ground. When the little bristly larvae hatch, they climb to the tops of flowers to hitch a ride from a bee. Once attached, they cling on until they arrive at the bee's nest, where they turn into fat little grubs and eat the honey the bee has stored up for its own young. I photographed the beetle, then let it go amid grasses and bird's foot trefoil below the croft. It immediately clambered in an ungainly manner under some plantain leaves out of the sunlight and seemed to start eating the trefoil.

I sent the photo to Bob Pope, a beetle expert at the British Museum of Natural History in London, whom I had known from many years ago in Sussex. Bob's reply was even more interesting. 'It seems to be an example of *Meloe violacetis*, Marsham. This species is less common than *Meloe proscarabaeus* (the common oil beetle), is rare in Scotland but has been taken from the Argyll area ... The common species occurs even further north, as far as the Moray Firth.' So I had made a first discovery of an important and interesting insect in Inverness-shire. I was delighted and ever since have taken specimens or photos of rare insects, butterflies and moths to the British Museum, registering a small number of first sightings for the region.

One blazing hot day, feeling guilty that I'd sunbathed all afternoon, I set off for a long trek over the hills to Deer Point and back along the hard way along the western shore, among boulders which dwarfed the mere human, still hoping I might find one of the dens the foxes were currently using. As I ascended the final hill below a vast flat, almost circular area filled with peat bog and floating banks of moss which formed the main rain-collecting area that fed my burn and three others, I saw a cuckoo behaving oddly. As I hid in the bracken it flew waveringly, like a sparrow-hawk might if it was dying of old age, fluttering low over the grasses as if checking nests in which to lay her eggs.

Suddenly a pair of meadow-pipits flew up, circled round her, looping and swooping excitedly but not actually buffeting the bigger bird. Then as they flew away again the cuckoo landed on a large mossy rock 50 yards away and sank down low as if not wishing to be seen. A female cuckoo spies on potential victims and has an egg ready to deposit roughly every 48 hours. She has to keep in her memory precise knowledge of the exact condition of several nests at a time. If she lays before the foster parents have laid their first egg, they may throw hers out or desert the nest. If she lays after the mother bird has started to incubate her own full clutch the other youngsters may hatch out faster and beat the young cuckoo in the competition for food. So, to find all the nests in which she may lay up to a dozen eggs in a season, the female cukoo does not waste time and energy searching the terrain on wing or foot – she perches in a tree and keeps watch on her victims. Once she has located a nest that way she flies down to inspect it. But here there were no trees and this cuckoo was using the mossy *rock* as a spy perch.

She waited there for about ten minutes while I took photos. Then, as if unable to hold back any longer, she flew direct to the place where the meadow-pipits had risen, vanished in the grasses for a few seconds and flew off again, disappearing behind a small ridge to the right. Keeping my eye on the exact place, I walked over and found the pipit's nest with four of their own reddish blotched eggs plus a slightly larger, rounder, paler and warmer egg – certainly the cuckoo's. Noting its position with an impromptu 'Treasure Island' map, and excited that I might later take photos of a young cuckoo ejecting its foster brethren or being fed, I went on my way.

As I turned west between the high hills above Deer Point the strong breeze from the sea turning my perspiration into the effective body cooler it's meant to be, I heard a spitting snarl, saw a flurry of reddish-brown and had a glimpse of blackened hairs at the tip of a thick brush as a big fox disappeared back into the gloom of a recess in the rocks. Just then the pungent sweet-ammoniac scent of the fox reached my nostrils.

Cursing myself for not going slower, at the sun being in my eyes and for my slow reaction because for once I actually had the camera and lens in my hand, I went over to look. It didn't appear to be a regular den, just a hole between the rocks of a small slide which vanished in a dark bend no more than a foot high. There were no scats about, no prey remains or signs of an oft-used trail, so it was probably one of many resting and watching places the fox used. Certainly there were some vole runs in the grassy edges of the crater-like area of bogs, and many frogs hopping round the pools, some of which were black with tadpoles.

I was sure now that the fox used this area for hunting, maybe more often than the big corrie above the western shore, and I tried the baiting technique once more. But though I waited out two more nights the meaty bones remained untouched. Had I seen Iain at this time I would have told him of the den I had found but it was some weeks before we met again and by then the lambs were bigger and the killings had stopped – possibly also due to the increase in natural prey. When I returned in ten days to photograph the young cuckoo in the meadow-pipits' nest I was doomed to disappointment. Everything had gone, nest and all. I was sure the fox had found it, probably using the nest as a convenient carrier bag to take away the eggs or youngsters.

Another small tragedy occurred a few days later. I woke up one rainy morning to guttural croaks from four ravens which had discovered a dead lamb caught up in some rocks lower down the burn and were making a meal of it. They appeared to be of one family, two adults and two youngsters, bulkier and loose plum-aged, with smaller heads and softer-looking about the eyes. Standing on the bank, the youngsters watched their parents with rapt attention as one stood on the lamb's body and appeared to lift its tail while the other pecked hard until it was able to tug some of the entrails out of the rectum. There was much croak-ing, and I had no doubt as I peered from my window – one move-ment of the door and the ravens would have flown – that the parents were actually teaching their young the technique which allows ravens to secure their carrion meal without having to puncture the dead animal's hide first. Once they have broken their fast on tender innards, the big birds can relax more and spend time attacking through the flank skin or through the front of the neck, which they are perfectly able to do. It was not the most pleasant sight in nature perhaps but interesting.

After a couple of hours the ravens flew away and I went to examine the lamb. It was not injured but was heavily swollen. As it had not come from the flock around the croft and was small for the time of the year, I concluded it had been late born, had not received much suck and had fallen into the water from the steep rock faces higher in the hills and had drowned. Then it had been washed down in the spates following recent rains. Oddly its eyes were still in place. I decided to leave it to see what other predators might come to the feast.

The torrential downpour accompanied by driving south-west gales continued throughout that day and the next night, and it became so cold that I had to light the paraffin stove. Even spiders and beetles were sneaking into the house to avoid the weather. Next morning the burn, now swelled to winter height, had washed the dead lamb a hundred yards out to sea, but when the wind switched to slightly warmer southerlies it drifted on shore again, this time on the east side of the bay and behind some rocks. Nothing went near it all day until, at about 7 p.m., the ravens arrived again, the original family plus three others. Failing to find the lamb where they had left it in the burn, they flew to the shore. A few were pacing about, and from the occasional croaks that came through my slightly open window they seemed to be talking to each other.

Presently one bird flew out over the waves, circled round the west side of the bay and went down again. I could not see the lamb from the window and wondered if it had been moved out to sea again, or had sunk. Then another adult raven flew out, this time in an easterly circle, baulked suddenly, gave a loud '*krok krok*' call and wheeled straight down to vanish behind some rocks. The other ravens, including the two young, needed no second bidding and flew to join in. It seemed the birds had worked out that if they sat on the rocks and just let one at a time search for the lamb they would find it without all tiring themselves any more than they need. Not until the following morning, when strangely the ravens failed to return, did a cruising greater black-backed gull find the carcass. After this big bird's visit I found the eyes had been taken. Not once had I seen a hooded crow near the lamb, though of course I had not watched it every minute of the day.

In mid-June large sunny patches of blue appeared in the sky, the winds dropped and summer was upon us once more. Within three days the bird life appeared to have doubled about the croft. Nesting willow-warblers sang sweet diminutive songs from above their young in the boggy thickets. Chaffinches, robins and wrens, also nesting, sang in their small territories, and a pair of yellow-hammers, rare for the area, had moved in from the south. One sang from a garden post for two mornings, his usual 'bread and no cheese' song ending here with a definite Highland accent so that it sounded like 'we'll let the de'il take ye'. Once even a bright yellow-green siskin appeared, a long way from its favourite conifer habitat. Perhaps it was after insects for its young as it competed with some fly-catching grey wagtails in fluttering gamboge dances near the cliff walls of my bay which, now the winds had dropped, was a trap of halcyon calm for thousands of insects. Butterflies such as small fritillaries, tortoiseshells and green-veined whites hopped and dropped through the warm humid air – infinitely preferable to the clouds of midges that had begun to bite in late May.

Now one of the buzzard pair, whose nest I could not find, and more often the male than the female, sailed over each morning, announcing its arrival with clear ringing '*keeyoos*' which allowed me time to grab my camera. After lunch one day I was at my desk when a flash of barred feathers made me look out. A superb little sparrowhawk with glaring eyes had landed on the garden post nearest the bird table. Briefly it stared in at me and in a trice was off again. But not before I'd seen its tail feathers were now perfectly grown, for I was sure it was Buzz.

No doubt he had wintered in Shona forest, keeping to the shelter of the glades where small woodland birds lived and sheltered too, until the better weather allowed him to hunt more in the open again. It was wonderful to know he had survived, that perhaps the time of year, the summer foliage, the sunlight and whole look of the place where he had learned to fly and hunt, had drawn him back so that this was now an integral part of his summer hunting area. Down on the shore Gilbert Gull was still occasionally flying to a rock to be fed a mussel or two, though he no longer allowed a close approach either.

As I thought of these old friends I remembered Harry Heron and how his old nest had been empty in the spring. I decided to check the whole heronry and see if he had mated this year after all. I boated to Deer Island and tramped beneath the trees where young herons squawked harshly '*ar ar . . . quarr.r.r. quarre*' their necks stretched up like thin grey bottles with button eyes near the tops. There seemed to be 7 new nests, yet a few of the older ones of the original 38 appeared to have been abandoned. Some of the adult herons were more wary than others and just kept sailing high over the treetops. A few flew in, landed heavily, spotted me and took off again, while others, with perhaps slightly younger offspring, came sailing in sneakily low and alighted gently. One heron sitting in the topmost twigs slightly to the side of a new nest stayed where it was, refusing to budge as I passed an open patch below. As I put the glass up the unmistakable broken toe stood out clearly. Certainly there were young in the nest but without climbing irons there was no way I could get up the tall fir. Besides, I consoled myself, it was far too late to start trying to build treetop hides.

On the nearby islets I took a few photos of young common gulls as they crouched in the grass. Four of the nests contained eggs in which half-formed young had died, some, it seemed, in the act of hacking their way out with the egg teeth on the tops of their little beaks. Five other eggs had not hatched at all. Out of a total of 14 nests this seemed a high mortality, probably due in part to marine pollution in the parents' food, for in the world's oceans pollution does not stay only in the areas where it is released but permeates in diluted form, theoretically, right around the globe. Perhaps the wildcat or a fox had kept the parents off the nests all night in the recent bad weather and the chicks in unhatched eggs had died of cold. The terns had not nested on the islets at all this year, though a few desultory individuals were flopping into the water offshore for fish. Perhaps it had been a bad winter for them in the Antarctic.

I rowed home in the late sunlight. Sammy the seal and two others sported and dipped near the brightly flashing oars, allowing the closest photos ever, and when I hauled the boat up they came close into my bay, treating me to a fine diving display as they constantly broke surface and sounded again with hissing snorts, the water falling and gleaming from their black polished pates.

There was an added bonus to that golden day for no sooner had I stepped into the croft than Little Fat Sergeant and his new family were back on the bird table. His wife was sitting on a twig weaving her tail sideways and he exchanged loving 'kisses' with her, their beaks together and gently plucking at each other's plumage, as if to help their summer moult. He not only fed the young, I noticed, but also popped the odd morsel of bread into her mouth as well. It made me feel quite lonesome.

19

Into a Last Wild Place

The thought that I would soon leave Shona for the even wilder place up the freshwater loch began to weigh on my mind. Certainly with its woodlands, far higher mountains where golden eagles soared and miles of uninhabited glens and rivers, it was a wonderful wildlife area. But could I really bear to live away from the sea? I would be going to far greater isolation. Even with the Land Rover parked in the pine wood I still faced a 13-mile return boat trip every time I needed supplies – easy enough in summer but hard in winter when the loch would often be a mass of seething foam.

One cloudy afternoon, when such doubts assailed me, I happened to look to the east and saw the only patch of blue sky in sight was now hovering over where my new home lay tucked below the hills. It was the first time I had ever seen that for usually my croft was bathed in sunlight while the cloud cover started a mile inland. How odd! In an isolated life such signs assume a strange importance.

Next morning dawned in a dull calm. Realizing I would need good cut timber at my new home, I decided to take the partly built cabin apart. As I set off down to the boat I thought I caught a glimpse of Buzz again, so I put a small chunk of steak near his old perch. At the site I dismantled the two walls with both sledge and claw hammers, jemmied up the 12-foot floor joists, removed all the nails and carried everything down to the rocks above high tide level. It was hard, sweaty work, and when I rowed home at teatime I felt ready for a good soapy bath in bowls outside before the sun went down.

As I walked up past the garden and plucked a lettuce I saw that the steak had disappeared. Surely it had been taken by Buzz, for Gilbert never came up to the croft any more, and it was too heavy for a small bird to remove. After my late lunch it was still calm outside, and in an hour it would be high tide. To hell with a bath then. I put engine and tank on the boat, went round to the site, loaded up, made a huge 'boom' of the biggest timbers with a couple of timber-hitched ropes, and towed the whole lot back to my bay in one go. As I enjoyed my belated bath by the side of the burn I felt pleased; I had moved almost the whole 12-foot cabin in one day – the hardest work I had done since I first arrived on Shona and struggled to make one room habitable.

On a trek in hot sunlight next day, I saw Harry Heron gliding into a small inlet near a disused log dump two miles along the south shore. As he did not associate my human figure emerging from the forest with the one he knew on my beach, I stalked him like a deer. As I peeped through bushes, quietly clearing a way for the lens between the leaves, the sun streamed upon him through a gap in the trees. He was bathed in light, his light blue uniform, smokey dark blue epaulettes and long black plumes transforming him into His Excellency as he stood there on his stilts, haughty and aware. Two camera clicks, and he heard them and was winging away, his body moving up and down with the effort.

I walked back over the hills and saw not a single deer. This didn't surprise me for at summer's height many of Shona's deer forded the narrow shallows on the north shore at low tide during the night and went up to the high tops on the mainland where they could still find good pasture and escape the biting flies. As I approached the croft in the dusk I saw a strange form moving to and fro just above the ground in the evening mist. Sometimes I could see it, sometimes not, and it swung like a pendulum on an invisible cord. It was a ghost moth, one of the swift moth family, the males of which fly like this when trying to locate a female. Its chrysalis is an oddity too, for when the imago is ready to emerge it wriggles up through the soil to the surface.

When Iain offered to give up a Sunday to help me move to the mainland, I asked him how much he wanted to be paid.

'Och, I'll leave that to your good self!' he said. 'Helping folk move is part of my work.'

I thanked him with relief for alone it would have been a two-to three-day task. On 2 July I began hauling the cabin timbers and a few spare fish boxes over to a boathouse on the mainland. The move had started.

In mid-July, I made a quick trip to London and, after seeing all the year's films developed, managed to part-exchange the unwieldy 1,000-mm mirror lens for a new 640-mm Novoflex which was far lighter. It also had a larger aperture, pistol grip focusing, and was ideal for wildlife work. Knowing that I would need a more powerful engine and a better boat for the winter gales at my new home, I bought a 20-hp Johnson outboard and also ordered, via a Mallaig firm, a 15 foot 6 inches WITH 450 fibreglass boat which had a small semi-cabin up front. My finances took a dive back to well below zero but it was better than suffering premature death. I was assured the double-hulled boat would remain afloat even when full of water.

When Iain called to look at a dead lamb I had found (which he diagnosed as suffering from pulpy liver) he spoke of something that had obviously been on his mind.

'Now you have a new engine, what will you do with the old one?'

I told him that I would probably use it for fishing in the old small boat up the new loch. He was silent for a while.

'Will ye be needing a fridge up there? You'll no' be making many trips to the store from that place?'

I said I had not possessed a fridge or had the use of electricity for over seven years and, as always, I would use a cold box in the new burn.

Iain coughed. 'Well it's just that I have this calor gas fridge we'll not be needing any more. We're getting a new deep freezer to put on to the generator. I was thinking . . .' He paused a moment, wondering how to put it. 'Now, your engine is worth fifty, sixty pounds and the fridge only about thirty. Would you take our fridge for your engine plus an adjustment in cash?' He gazed at me intently, a slight smile on his face.

I looked at him sitting there cross-legged, a man I had misjudged on earlier meetings. He may have regarded himself as a servant, but to me he was a man of uncommon skills – bilingual (for he had the Gaelic), skilful fisherman, deer stalker, tree feller, sawmill operator, ferryman, gardener and shepherd. He was also a fine enough piper to have earned his living at the pipes alone. Although I had not wanted casual walk-in company while writing in the lonely cottage, I had come to enjoy his occasional visits and lively talk in the lambing seasons. I could see he wanted my 7-hp engine badly.

'No,' I said. His face fell.

'We'll do an Indian trade,' I said.

'What's that?'

'You want engine. I need fridge. Your need. My need. Straight exchange. No cash adjustment. That's it.'

Iain found it hard to believe but that was the final arrangement.

After he'd gone I carried the dead lamb down to the shore a quarter mile from the croft, set up a small hide in the bracken two hundred yards away and entered it before dawn. I still wanted to see which predator came first. Within an hour, and just after first light, a pair of greater black-backed gulls sailed in and for two hours I watched while they had the carcass to themselves. No ravens or hooded crows came near. When I went down to bury the remains of the lamb, the upper eye was missing and the powerful beaks of the gulls had hacked a hole below the ribs where they had feasted deep on lungs and heart. This again did not vindicate the crow completely, especially on inland hill farms where there are few black-backs.

*   *   *

The wildlife adventures during my remaining time on Shona became fragmentary as I worked to complete yet another 'final' version of the Canadian wilderness book. I was also repairing the croft, to fulfil my end of the deal before I left. I was installing new front guttering when Gilbert the gull came circling over and '*qwuckqwuck*'-ing with another herring-gull which kept at a greater distance. It seemed as if he might have found a mate, though I doubted they would stay together until the following year. He had grown most of his new wing feathers and I raced indoors to fetch my camera, managing a good picture of him in flight. At dusk Harry Heron flew into the bay accompanied by two youngsters. They kept together on the dry beach with their heads hunched into their shoulders, watching intently as he stalked and waited, stalked and waited, in the shallows. I was sure he was teaching them how to fish. Why in heck didn't he ever come back in photographable daylight!

Next day, when I boated back with a huge scythe borrowed from Iain for cutting the bracken, I saw a red deer hind actually climbing a fence. One stob had snapped off so that the fence was leaning away from her. She climbed slowly up it like a long-legged monkey, her long neck and almost reptilian head extended, putting one foot carefully above the other. She seemed to hook the wires under the two upper rear claws, then sprang off and away. Later, after cutting a quarter acre of bracken, I started sharpening the scythe while standing on the steep slope. Flailing away with the carborundum stone in a most professional manner, feeling pleased I was carrying out my policy for wilderness living – live quietly, enhance the natural beauties, pay the place back for what it has given you, leave no permanent fixtures and, like a nomadic Indian, depart without trace – when I felt a sudden sharp pain. Careless in self-congratulation, I had nearly severed my right index finger on the curved blade of the scythe and the blood spurted. Afraid that the deep cut would need stitches, I raced indoors for my animal medical supplies. I sucked more blood out hard, forced the top of the finger back down, ringed it with antibiotic cream, wrapped a bandage round and let the blood congeal inside it, so forming a hard protective cast. Fortunately the flesh knitted back together but typing was impossible for more than a week. I didn't like the scythe after that and found that by swinging an old saw flatways I could hack the bracken just as well, for I cut it on both the back and forward swings.

Two days of gales followed, raging so hard that leafy twigs snapped off the ash trees and were plastered against the wet windows, reminding me that autumn was not far away. When the third morning came in a cloudless blue sky, I boated for supplies to the village at the head of the loch five miles away, so that I could fish on the way. I was just trolling along when suddenly the rod felt strange, lighter, and the line was right up in the air by my side. My heart chilled, for I was just passing Castle Tioram and for a moment I thought that someone, maybe a ghost of the Clanranald chiefs, was sitting behind me and holding it up.

Looking back, I saw a tern had taken the lure and was flying at the end of the line like a kite, diving and swooping as it tried to get off. It weighed so little that it did not undo the rachet on the reel. I had been going a bit too fast for the weight I was using on the line and the artificial lure had been travelling near the surface when the tern, believing it to be a fish, had dived upon it. I slowed the engine, reeled in and the poor bird hit the water by the side of the boat. Luckily the hook was lodged only in the side of its beak and there were no flesh wounds. I removed it as the tern pecked hard with its sharp bill and let it go, none the worse for its odd adventure.

While I cooked supper that evening I heard the rutting roar of a stag behind the croft. I slipped out in camouflage jacket and stalked near as it roared again, I was sure it was Sebastian back from his summer on the mainland. His antlers were magnificent but they still had only ten points. I roared back, but after a few paces forward he turned and ran off up the hill. It was strange that he should be making his rutting roars as early as mid-August, especially when there were apparently no hinds near, a fact I

confirmed after an early trek next day. Maybe he was just assert-
ing himself again in his old rutting territory.

When I returned from the walk two glowing garden tiger
moths, just hatched out, were resting in the sun on one of my
garden posts. A new hawker dragonfly, the black and yellow
ringed species of the *Aeshnas*, which a hasty look at my insect
books told me was 'not found in Scotland', was now working
above the burn for insects to snap up in flight.

I checked the few bramble bushes for blackberries, hoping that
some would be ready in time for bottling and taking to
Wildernesse, but the fruits were still green and red. The second-
year shoots of the big bush nearest the croft that had borne such
luscious berries last year had all died off and were brown and
sere, and the few new shoots growing through them had no flowers. Yet further up the burn the smaller bush which had yielded a
poor harvest last year was now filled with unripe berries. It was
interesting that brambles thus stagger their fruiting peaks, being
more prolific every second year, as well as each bush having ripe
and unripe berries from early September to mid-November. Thus
both humans and birds had a constant supply over at least a two-
month period, and with birds scattering seeds via their droppings, more new bushes could take root. I kept the brambles
trimmed, not only so the strength of the thick probing leaders
would go into the fruit but because they were dangerous to weakened sheep in winter.

It was a hot day and after lunch I sunbathed on the green
sward below the croft. The yellow stinging-nettle flowers were
exploding with little puffs of pollen into the warm air like
cannonfire, providing their own miniature Waterloo before my
half-closed eyes. When the sun sank lower I spent the rest of the
day forking rocks and large stones from a fifty-yard strip of
beach. On this cleared shingle I would berth my new boat in a
special rope and plank cradle I had made for her.

Towards the end of August she arrived at the little pier, and
she looked a beautiful creature. Digby Vane and Iain and some of

the other islanders and mainland friends were also there as we
lowered her into the sea. She sat like a swan, high, light and buoyant, as if basking in the admiring compliments. But I was in a
hurry to get her home at the peak of the high tide for I had to
haul her out alone and she weighed 400 lbs without the engine.
As I towed her back to my bay curlews with a few young were
back on the sandbanks from breeding further inland and several
cormorants passed across the bows bobbing their heads in and
out of the water as they tried to paddle out of range before
making a frantic last-second dive.

I hauled the first boat out as usual but found I was hardly able
to budge the new one. Only by splashing water to make the
planks slippery and by hauling her at a rush from the sea did I
manage to get her on to the cradle. Even then I pulled my back
slightly. When I looked down from the slope they both lay as if
talking to each other. It's just as well boats can't talk, I thought,
or the old one would tell the newcomer yarns about the Highland
storms to frighten her out of her wits.

During three drizzly days I pounded away at my book, but
when sun blazed on the fourth I decided to try out the new boat.
I carried down and screwed on the new engine which, with full
tank, brought the total weight of the craft beneath me to well
over 500 lbs. She moved superbly, solid but easy in the water, and
when I turned up the throttle and she planed at 22 knots I soon
slowed down, scared stiff at the unaccustomed speed. I always try
to go quietly in a boat but it was good to know the power was
there if needed up the long loch that was soon to be my home
water.

So delighted was I that I went further than I had intended and
burbled on slowly past Deer Island. To my chagrin there were 48
herons sitting on the rocks below the great nests. The dark slabs
were thick with both adults and youngsters, all sitting hunched
like flecks of snow in the dying sunshine, and I had no camera
with me! By the time I returned the tide was well past the ebb.
Boat and engine were now too heavy for me to haul up the steep

banks of shingle. So, laying down planks and putting rocks on their ends to hold them down when the tide rose, I bent a long branch of the nearest alder tree down about eight feet and tied the boat painter to it. As the tide came up and floated the boat the spring-back of the branch would constantly haul it inch by inch out of the sea's reach. The idea worked too, as I found when I went down at midnight and made the final heave to safety.

On the last day of August the worst gale since winter came blasting in from the south-west. Anxious to fetch supplies so that I could spend a final week on the Canadian book before the move, I set off in the new boat. After a mile I was overtaken by pounding rain, the gale force increased and I had to anchor in the lee of Castle Tioram rather than at the pier, which was a mass of raging foam. On the way back the boat performed feats of which the old dinghy would have been incapable. Her bow drove deep into each crashing wave and reared again easily, the semi-cabin diverting water to each side with cleaving thrusts so that I no longer had to contend with drenching spray, just the rain alone. Even so I had to dash into the lee of the little inlet opposite Ballindona. My experience told me that the gales would now increase for several hours before lessening and after waiting several minutes I made a dash for it. Although a little water was shipped from the high waves, the powerful new engine thrust us home in minutes, smashing the way ahead. With the tide now low I left her on the sand with the double disc anchors out from her stern. Then I tied her to the alder with over a hundred yards of new, steam-tarred nylon rope.

I was just in time. Within minutes of being back in the croft I could hardly even see the land opposite, and the sea had turned into a mass of raging white. Crags and grey prominences appeared occasionally through the swathes of hissing rain and mist, while the wind howled through the ash trees, tearing off branches, rattling my roof, and bent bushes almost double on the slopes below. Despite the comparative shelter of the cliff-walled bay the rising tide came in with great crashing waves, so that the

new boat dragged its anchors and the bow rope became snagged
round several weed-covered boulders. It was now impossible to
haul it in straight. As I fought for footing, up to my waist in the
waves, trying to free the rope again, the boat pounded up and
down at the edge of the tide.

Somehow I got the engine off and the fuel tank and oars out
before the boat began hitting the small rocks of the middle beach.
Terrified the boat might yet be damaged I made superhuman
efforts and managed to haul it out over planks. I was down there
for three hours and, August or not, found myself trembling with
fatigue and cold. It was just like the day I had lost my boat below
the clifftop cabin in Canada and I really felt that some strange
wilderness nemesis, which appeared whenever I was about to
leave a place, had again overtaken me. That too had been a
summer storm. After seven years of living with small boats and
the sea it seemed I could still make mistakes and misjudge the
weather.

Up the slope the swollen burn had burst its banks, flattening
and tearing off some nettles. Several tortoiseshell butterfly caterpillars lay half drowned in the ooze. I warmed them up in my
hands before placing them on new nettles in the shelter of the big
rock below my window. In a minute they all recovered, and with
their great rounded clasper feet set firmly they clipped away at
the leaves like little machines thrust into high gear.

This sudden touch of harsh weather seemed to remind the
birds that winter was no longer far away and many began flock-
ing to the bird table – young robins, hedge-sparrows, chaffinches,
blackbirds, all with one parent or the other. Wren families flitted
in the bracken and once a young wren landed on the bird table
and squared up to the hen chaffinch, who ignored it. Harry and
his two youngsters came to the bay nearly every evening now,
even when I was collecting cockles and mussels to bottle. On 4
September a young buzzard landed on the rock outside the
window. I had never before seen one so close. It looked like a
small eagle but with brown eyes and a softer expression. When it

flew away I put a piece of meat on the rock, hoping it would return. Within the hour there was a flash across the window, the sound of rushing air, but it was not the buzzard. Sparrowhawk Buzz snaffled the meat and was away again so fast I was up in time to see only his long tail vanishing into the cliff bushes to the south-east. His speed was now dazzling but it was good to know he was still alive.

I finished the new Canadian book on 12 September, eventually calling it *Alone In The Wilderness*. Brilliant sunshine blazed during those last three days as I carried gear down to the beach – fishbox cupboards, pictures, sea-scoured ornaments from the roots of trees, ten times more books than I had first brought with me, the wooden scaffold for painting the roof, and tools, boots, tinned food and clothes – covering all with plastic sheets held down by rocks in case the weather changed again. I dug up all the onions, carrots and cabbages, bagging some and brining others.

One evening, as I returned from stowing two more loads into the boatshed on the mainland and came up the slope in the dusk, the leading hind of a group of nine that were grazing behind the croft barked at me like a bear. I stayed still and waited. Sure enough, Sebastian was with the hinds and later I heard him roaring. A real master stag now, he was having first choice of the hinds and would keep all young rivals away until he'd had his fill of rutting.

Harry Heron came floating alone into the bay early in the morning, making his familiar harsh '*kraink kraink*' calls, sounding like someone tearing a heavy sail. I went out with the camera to call to him, wondering if he would take any notice. By constricting my throat I made a passable imitation. To my surprise he immediately wheeled round and flew towards me. It is quite a frightening sight to see such a huge bird, dagger beak extended, coming straight for you, but I kept still and darned if he didn't land on a large rock just a few yards from the cottage. I took his picture, then he flew away again. It was perhaps the most astonishing experience I had with him in his maturity.

On the last day I tore up the old blue carpet and hauled it with the last items down to the beach – more books, the ancient paraffin heater, the desk contents, clothes, tape recorder, little battery record player, containers, cold box, and the old fish net to protect my new garden patch. I took down everything but my bedding, the cooker and the big mahogany desk I had grown to love and was now loath to leave behind.

As I worked a lovely red admiral butterfly, the first I had seen at Ballindona, landed on some white bramble flowers. Dandelion seeds were blowing in the wind on their tiny parasols, craneflies were laying eggs in the earth, and the first blackberries had ripened on the bushes, moving slightly in the wind like farewell offerings.

In late afternoon Iain came round with the island's largest dinghy. We loaded the three boats and took everything over to the mainland woodshed. There my big boat and new engine were lifted on to the trailer of a friend who would leave them on the new lochside where the little farm track ran out six-and-ahalf miles from Wildernesse. Personal items were loaded into my Land Rover and everything else on to a flatbed coal lorry which Iain was to drive on the morrow and which had been lent me by Allan MacColl.

On the way back to Ballindona for the last time I called at Shona House for a farewell drink with the Vanes. I had dreaded this moment most of all. As I sat in their lovely lounge, my hand none too steady on the huge dram of whisky Digby handed to me, I tried to thank them for letting me live on their island and for the loan of the radio, the disc anchors and the fine mahogany 'J. M. Barrie' desk which I had grown to love as if it were an animate thing and which seemed to have been 'lucky' for my writing.

Suddenly Kay Vane said: 'You'll need that desk now, Mike, so you had better keep it, with our good wishes.'

Reggie Rotheroe coughed lightly and spoke of the radio which he had made himself. As everyone on the island had his little radio mine was now surplus to requirements, so I had better keep that too.

'Don't bother about returning the anchors or the calor gas regulator,' said Digby. 'You'll need them up there, and you'd better take this along as well.' He reached behind his chair and produced a new five-gallon water container with a tap. 'We never use it and it may come in useful for your drinking water in a winter freeze-up.'

Don't worry if life doesn't work out too well up there,' said Kay. 'We've talked things over and you can come back here any time you wish. We'll keep Ballindona free for you to return to for at least two years.'

I stared at them, scarcely able to stammer out my thanks. I think I hid my emotions rather well. They had been more than good to me, these people.

It was dark when I boated back to the little beach and walked up the steep slope for the last awful night. I recalled then the last day at my clifftop home in Canada, when useful objects had floated by on the evening tide as if begging me not to go. Now I was saying farewell to this wild, beautiful place with the same reluctance.

With everything gone, I sat staring into space. The washing mirror gleamed its bleak emptiness on the wall, reflecting the empty shelves, the bare wooden floor, my stew simmering on the little stove. I felt a strange emptiness and everything sounded louder, as if I were sitting in a drum, a void. A few house flies swung silently to and fro in the still air as if they had moved in for the wake. It was hard to sleep for the last time in the little bed on the fishboxes.

Early next morning, with the weather finally broken and rain teeming down, I said my goodbyes to Ballindona, to the burn, to the ash trees, to the heathered hills, to the room where Buzz had spent his nights during the weeks of training, where Gilbert the gull had recovered his health, and to the croft itself. I walked down the path with the final load, oblivious to the tears of the sky, and set sail in the old boat for the last time. Sparrowhawk Buzz appeared from nowhere and flew low over my head as if assuring me he was well and would survive all right. Before I had gone half a mile Harry Heron, looking immense against the dark and raining sky, also flew over with terrible harsh cries. Herons carry away the souls of the departed, the ancient Highlanders believed, but as he came really close I felt that he sensed I was leaving, was even saying goodbye.

I hated to go, yet leave I must, to follow the lonely star of my fate to even greater isolation and deeper Highland wilderness experience. It was not fear I felt, for loneliness had for me become a liberation, but there was more than rain upon my cheeks as Ballindona receded in the mist for I knew now that while Canada had been a step aside – a rebirth of a kind – on Shona I had found a personal meaning to life itself.

Also available by Mike Tomkies

A Last Wild Place
Seasons in the Wilderness

When Mike Tomkies moved to a remote cottage on the shores of Loch Shiel in the West Highlands of Scotland, he found a place which was to provide him with the most profound wilderness experience of his life. Accessible only by boat, the cottage he renamed 'Wildernesse' was to be his home for many years, which he shared with his beloved German Shepherd, Moobli.

Centred on different landscape elements – loch, woodlands and mountains – Tomkies describes the whole cycle of nature through the seasons in a harsh and testing environment of unrivalled beauty. Vivid colours and sounds fill the pages – exotic wild orchids, the roar of rutting stags, the territorial movements of foxes, otters and badgers, an oak tree being torn apart by hurricane-force gales. Nothing escapes his penetrating eye.

His extraordinary insights into the wildlife that shared his otherwise empty territory were not gained without perseverance in the face of perilous hazards, and the difficulties and challenges of life in the wilderness are a key part of this remarkable book.

ISBN 978 1 78027 703 5